PRACTICAL
SEAL DESIGN

MECHANICAL ENGINEERING

A Series of Textbooks and Reference Books

EDITORS

L. L. FAULKNER

Department of Mechanical Engineering
The Ohio State University
Columbus, Ohio

S. B. MENKES

Department of Mechanical Engineering
The City College of the
City University of New York
New York, New York

OTHER VOLUMES IN PREPARATION

PRACTICAL
SEAL DESIGN

LEONARD J. MARTINI
Naval Ocean Systems Center
San Diego, California

MARCEL DEKKER, INC.   New York and Basel

Dedicated to my Dad and Mom

Library of Congress Cataloging in Publication Data

Martini, Leonard J., [date]
    Practical seal design.

    (Mechanical engineering ; 29)
    Includes bibliographical references and index.
    1. Sealing (Technology) 2. Elastomers. I. Title.
II. Series.
TJ246.M37 1984   621.8'85   83-26306
ISBN 0-8247-7166-4

MARCEL DEKKER, INC.
270 Madison Avenue, New York, New York 10016

Current printing (last digit):
10 9 8 7 6 5 4 3 2 1

PRINTED IN THE UNITED STATES OF AMERICA

Preface

This book was written to help fill a gap in the mechanical engineering literature on elastomeric seals. Up to this time, designers have had to depend on O-ring seal handbooks and product sale manuals, which at best are subjective and limited in theoretical and practical application. *Practical Seal Design* provides the designer with a comprehensive discussion of the theory and practical application of elastomeric ring seals. The theory is reduced to practical use and presented in a manner conducive to solving current sealing problems. Part I of the book is devoted to a general discussion of elastomeric ring seals, including specific topics in elastomeric ring seal geometry, material-compound capability, and material performance under various environments. Part II is devoted to the detail of specific applications of static, reciprocating, and rotary seal function.

Although the book contains various sections of theoretical discussion, it has been written with the designer in mind. General and specific design methods are presented together with solved engineering problems that elaborate important design considerations. In this regard, the book is helpful to students who want to understand seal theory and learn to apply it to actual field problems. The book centers around the use of elastomeric O-rings, but the design methods and practical engineering considerations are applicable to most other types of elastomeric seal configurations.

The book is full of aids for the seal designer. O-ring specifications for military and aerospace standards, tube fittings, and electrical connectors are consolidated for easy access. Property comparison tables and temperature capability charts for the currently used elastomeric compounds are presented. To aid the

designer in trade-off studies, a series of material performance
charts for selecting the proper elastomeric material based on temperature, environmental, and physical criteria is provided. A list
of some 49 elastomeric ring seal manufacturers and distributors
are presented arranged by location. Detailed design considerations, such as O-ring stretch, swell, shrinkage, and blowout prevention, are culminated with a designer's quick reference table
for important design criteria factors.

Under specific applications of O-ring seals, a chapter is devoted to clearing up the confusion between military and industrial
specifications for O-ring gland designs, together with design example problems. Military and industrial bases for tube fittings
are discussed in Chapter 4, and equation methods and compression
load charts for face-seal glands are also presented. Chapter 5
forwards a theoretical method for determining the effect of side
loads on a piston within a cylinder and the restoring force required
for concentricity. Nomograms for easily determining O-ring friction as a function of pressure and cross-sectional squeeze for
cylinder-piston application are presented, together with a discussion on minimizing system hysteresis. Each of these subjects is
followed by design examples. Chapter 6 is devoted to rotary seal
design for shafts using standard elastomeric O-rings. A revolu-
tionary method of extending the life of O-rings in rotary seal applications is analyzed in detail. The method results in a seal gland
design that puts the O-ring in peripheral compression to counter-
act frictional effects produced by the shaft running against the
O-ring seal. Slanting the O-ring glands relative to the axis of the
shaft further extends the life of the seal. The chapter concludes
with practical design data for manufacturing such rotary seal
glands and design tables for the practical designer. Although the
design tables cover the entire range of O-ring sizes, the designer
is cautioned because only a few of the sizes have been verified by
actual hardware. Those rotary shaft seal designs that have been
reduced to practical use have proven successful for many years.

The book attempts to cover the practical applications of elastomeric O-ring seals for the designer who must make responsible engineering decisions. Although most of the theoretical concepts
presented in this book are reduced to practical application and el-
aborated by design example problems, the designer is cautioned to
use discretion when applying the detailed concepts to particular
design problem. The ancient commendation of wisdom by King
Solomon is appropriate:

I, wisdom, dwell with prudence
And I find knowldege and discretion
                    (Proverbs 8:12)

The experienced designer has learned to use discretion in all judgments, and verifies designs through actual testing whenever possible. This is particularly true in the field of seal design.

The author wishes to express his appreciation to those who helped in the preparation of this book: secretaries Joan Goddard, Margret Cole, and Betty Kimberly; engineer in training, Mike Phillips; and the many publishers and corporations for their permission in using the various materials and information incorporated in the writing of this book. I also wish to thank the staff of Marcel Dekker, Inc., for their help and encouragement. This book has grown out of my previously published work Designing for Elastomeric Ring Seals, a chapter in *Plastic Products Design Handbook,* edited by E. Miller, Marcel Dekker, Inc., 1981.

Leonard J. Martini

Contents

Figures, Tables, Charts, and Design Examples

Figures

Tables

Charts

Design Examples

PRACTICAL
SEAL DESIGN

1
ELASTOMERIC RING SEALS

1

Basic Configuration of Elastomeric Ring Seals

Elastomeric ring seals are circular rings of various cross-sectional configurations installed in a gland to close off a passageway and prevent escape or loss of a fluid or gas. Designing for elastomeric ring seals depends on three major and interrelated variables: the operating conditions or environment the seal will experience, the gland geometry into which the seal will be installed, and the seal material and geometry. The various interrelations of these three variables account for the fact that there are so many different types of seals and applications.

Figure 1 shows a cross-sectional view of an oil pump incorporating various common applications of static and dynamic elastomeric ring seals. Static seals do not see relative motion between themselves and the parts they seal. In Fig. 1, O-rings are used on the inlet and outlet tube fittings and top and bottom flange plates, these being static seal applications. Other static seal applications include face seals and seals for electrical connectors. Dynamic seals experience relative motion between themselves and the parts they seal, typically used on pistons, rotating shafts and intermittent face seal applications, as in check valves.

This chapter presents for easy reference the various kinds of elastomeric seals currently available to the designer: that is, cross-sectional configurations, sizes, and materials. Properties of the commonly used elastomers are discussed and finally summarized.

I. GEOMETRY AND APPLICATION

There are at least 49 American manufacturers of some 20 different types of elastomeric ring seals. (A list of manufacturers and dis-

3

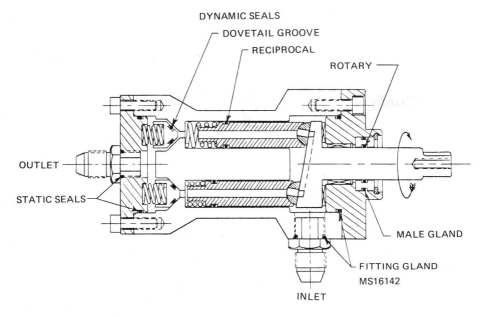

Figure 1. Elastomeric seal applications

tributors appears in Table 8.) Table 1 presents 12 basic confiura-
tions and their applications, most other types being derivatives of
these. The circular cross-sectioned seal, commonly known as the
O-ring, is the most used and the least expensive. It is used in
two general design applications: static and dynamic. Static applications may range from vacuum to over 60,000 psi for sealing
flanges and O-ring grooves. One such seal is the Bal-Seal made
for static face seals. Design manuals and catalogs are usually
available free of charge from all estomeric seal manufacturers.
Such information should always be consulted when designing
glands for uncommon types of seals and their application.

   Elastomeric seals with lobed cross sections were designed for
both static and dynamic applications. The four- and six-lobed
configurations resist spiral failure and also extrusion failure in
applications with large clearance between parts.

   Of the some 20 different types of elastomeric ring seals available, the common O-ring type is the most versatile. The conventional type of O-ring may be used in almost any application if the
gland to contain the O-ring is designed correctly and the right
size and material is chosen for the O-ring.

Table 1. Elastomeric Ring Seals

Configuration/ cross section	Application/usage
O-ring	Static seal for flanges, face seals, fittings, electrical connectors Dynamic seal for limited speed and pressure of reciprocating and rotary sealing
O-ring with straight and contoured backup rings	Static seal for very-high-pressure application which may otherwise cause O-ring extrusion (for pressure greater than 1000 psi when total diametral clearance is greater than 0.010 in.) Dynamic seal to prevent spiral failure in reciprocating applications and extrusion in high-pressure reciprocating and rotary applications
Parker Polypak	Static, reciprocating, and rotary applications where pressures range from 800 to 60,000 psi; effective in out-of-round or tapered cylinder bores
SCL seal	Reciprocating rod-cylinder application to prevent the usual slight weepage seen in O-ring-type rod-cylinder applications; normal high pressure limit of 10,000 psi when used with backup rings.
U- or V-seals	Reciprocating rod and piston-cylinder applications; normal high-pressure limit of 3000 psi with backup rings.

Table 1. (continued)

Configuration/ cross section	Application/usage
 Piston Rod T-seals	Reciprocating: piston-cylinder and rod-cylinder seals to a limiting pressure of 10,000 psi without rolling or spiral failure; recommended for high-pressure long-stroke cylinder applications; clearance between piston or rod and cylinder wall may be increased beyond that recommended for conventional O-ring seals.
 Spring-loaded BAL-seal	Reciprocating rod and piston-cylinder applications, 0 to above 10,000 psi with less friction than conventional type O-rings (0.02 coefficient of friction) Rotary applications for reduced friction; limited to maximum PV = 75,000 psi-fpm[a] Static applications, 0 to 10,000 psi
 Spring-loaded static face seal—BAL seal	Static face seal, vacuum to 3000 psi, −70 to 550°F; may be used in standard O-ring grooves
 Piston cups	Piston seals for pressures to 10,000 psi, and rough cylinder bore finishes up to 250 rms, or where clearance between the piston and bore is wider than recommended for standard O-rings.
 Quad-X ring	Static seal for reliable performance at pressures beyond 2000 psi Reciprocating applications to resist and eliminate spiral failure Rotary applications for higher speeds than conventional O-rings

Table 1.   (continued)

Configuration/ cross section	Application/usage
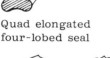Quad elongated four-lobed seal	Static and dynamic applications where large diametral clearances may otherwise result in extrusion of conventional O-rings Reciprocating applications to eliminate spiral failure
Piston Rod Kapseal	Reciprocating applications where low breakaway friction is required; this Teflon cap used in conjunction with an O-ring; sizes smaller than 2-in. diameter require a split-groove housing

[a]PV is the pressure (psi) exerted on the O-ring X the shaft velocity (fpm).

This book presents information relative to standard elastomeric O-rings, but most of the design procedures and general design considerations can be applied to almost any elastomeric seal configuration. The sections on compounds and materials are of course, independent of cross-sectional configuration and applicable to other cross-sectional types.

II. SIZE

Elastomeric O-ring seals have been standardized under the basic industrial standard dimensions of AS568, Aerospace Standard published by the Society of Automotive Engineers, and a multitude of military standards; AN6227 and MS28775 for general use; M25988, M83248, MS9020, MS9355, and MS29512 for straight-thread tube fittings; and MS28900 for electrical connectors. Most of the major O-ring manufacturers produce all the sizes specified in industrial standard AS568, as presented on Tables 13 and 22. Comparisons between the industrial and military standards are presented and discussed in Chap. 3.

AN6227 (Table 2) is an Air Force-Navy Aeronautical Standard that covers 88 sizes of AS568. MS28775 (Table 3) is the basic

Table 2. Air Force-Navy Aeronautical Standard, Hydraulic O-Rings Packings

FSC 5330

ENLARGED VIEW OF SECTION A-A

DASH NO.	NOMINAL W	NOMINAL ID	NOMINAL OD	ACTUAL SIZE W	ACTUAL SIZE ID	DASH NO.	NOMINAL W	NOMINAL ID	NOMINAL OD	ACTUAL SIZE W	ACTUAL SIZE ID
1	1/16	1/8	1/4	.070 ±.003	.114 ±.005	46	3/16	3-3/4	4-1/8	.210 ±.005	3.725 ±.015
2	1/16	5/32	9/32	.070 ±.003	.145 ±.005	47	3/16	3-7/8	4-1/4	.210 ±.005	3.850 ±.015
3	1/16	3/16	5/16	.070 ±.003	.176 ±.005	48	3/16	4	4-3/8	.210 ±.005	3.975 ±.015
4	1/16	7/32	11/32	.070 ±.003	.208 ±.005	49	3/16	4-1/8	4-1/2	.210 ±.005	4.100 ±.015
5	1/16	1/4	3/8	.070 ±.003	.239 ±.005	50	3/16	4-1/4	4-5/8	.210 ±.005	4.225 ±.015
6	1/16	5/16	7/16	.070 ±.003	.301 ±.005	51	3/16	4-3/8	4-3/4	.210 ±.005	4.350 ±.015
7	1/16	3/8	1/2	.070 ±.003	.364 ±.005	52	3/16	4-1/2	4-7/8	.210 ±.005	4.475 ±.015
8	3/32	3/8	9/16	.103 ±.003	.362 ±.005	53	1/4	4-5/8	5-1/8	.275 ±.006	4.600 ±.015
9	3/32	7/16	5/8	.103 ±.003	.424 ±.005	54	1/4	4-3/4	5-1/4	.275 ±.006	4.725 ±.015
10	3/32	1/2	11/16	.103 ±.003	.487 ±.005	55	1/4	4-7/8	5-3/8	.275 ±.006	4.850 ±.015
11	3/32	9/16	3/4	.103 ±.003	.549 ±.005	56	1/4	5	5-1/2	.275 ±.006	4.975 ±.015
12	3/32	5/8	13/16	.103 ±.003	.612 ±.005	57	1/4	5-1/8	5-5/8	.275 ±.006	5.100 ±.023
13	3/32	11/16	7/8	.103 ±.003	.674 ±.005	58	1/4	5-1/4	5-3/4	.275 ±.006	5.225 ±.023
14	3/32	3/4	15/16	.103 ±.003	.737 ±.005	59	1/4	5-3/8	5-7/8	.275 ±.006	5.350 ±.023
15	1/8	3/4	1	.139 ±.004	.734 ±.006	60	1/4	5-1/2	6	.275 ±.006	5.475 ±.023
16	1/8	13/16	1-1/16	.139 ±.004	.796 ±.006	61	1/4	5-5/8	6-1/8	.275 ±.006	5.600 ±.023
17	1/8	7/8	1-1/8	.139 ±.004	.859 ±.006	62	1/4	5-3/4	6-1/4	.275 ±.006	5.725 ±.023
18	1/8	15/16	1-3/16	.139 ±.004	.921 ±.006	63	1/4	5-7/8	6-3/8	.275 ±.006	5.850 ±.023
19	1/8	1	1-1/4	.139 ±.004	.984 ±.006	64	1/4	6	6-1/2	.275 ±.006	5.975 ±.023
20	1/8	1-1/16	1-5/16	.139 ±.004	1.046 ±.006	65	1/4	6-1/8	6-3/4	.275 ±.006	6.225 ±.023
21	1/8	1-1/8	1-3/8	.139 ±.004	1.109 ±.006	66	1/4	6-1/2	7	.275 ±.006	6.475 ±.023
22	1/8	1-3/16	1-7/16	.139 ±.004	1.171 ±.006	67	1/4	6-3/4	7-1/4	.275 ±.006	6.725 ±.023
23	1/8	1-1/4	1-1/2	.139 ±.004	1.234 ±.006	68	1/4	7	7-1/2	.275 ±.006	6.975 ±.023
24	1/8	1-5/16	1-9/16	.139 ±.004	1.296 ±.006	69	1/4	7-1/4	7-3/4	.275 ±.006	7.225 ±.030
25	1/8	1-3/8	1-5/8	.139 ±.004	1.359 ±.006	70	1/4	7-1/2	8	.275 ±.006	7.475 ±.030
26	1/8	1-7/16	1-11/16	.139 ±.004	1.421 ±.006	71	1/4	7-3/4	8-1/4	.275 ±.006	7.725 ±.030
27	1/8	1-1/2	1-3/4	.139 ±.004	1.484 ±.006	72	1/4	8	8-1/2	.275 ±.006	7.975 ±.030
28	3/16	1-1/2	1-7/8	.210 ±.005	1.475 ±.010	73	1/4	8-1/2	9	.275 ±.006	8.475 ±.030
29	3/16	1-5/8	2	.210 ±.005	1.600 ±.010	74	1/4	9	9-1/2	.275 ±.006	8.975 ±.030
30	3/16	1-3/4	2-1/8	.210 ±.005	1.725 ±.010	75	1/4	9-1/2	10	.275 ±.006	9.475 ±.030
31	3/16	1-7/8	2-1/4	.210 ±.005	1.850 ±.010	76	1/4	10	10-1/2	.275 ±.006	9.975 ±.030
32	3/16	2	2-3/8	.210 ±.005	1.975 ±.010	77	1/4	10-1/2	11	.275 ±.006	10.475 ±.030
33	3/16	2-1/8	2-1/2	.210 ±.005	2.100 ±.010	78	1/4	11	11-1/2	.275 ±.006	10.975 ±.030
34	3/16	2-1/4	2-5/8	.210 ±.005	2.225 ±.010	79	1/4	11-1/2	12	.275 ±.006	11.475 ±.030
35	3/16	2-3/8	2-3/4	.210 ±.005	2.350 ±.010	80	1/4	12	12-1/2	.275 ±.006	11.975 ±.030
36	3/16	2-1/2	2-7/8	.210 ±.005	2.475 ±.010	81	1/4	12-1/2	13	.275 ±.006	12.475 ±.030
37	3/16	2-5/8	3	.210 ±.005	2.600 ±.010	82	1/4	13	13-1/2	.275 ±.006	12.975 ±.030
38	3/16	2-3/4	3-1/8	.210 ±.005	2.725 ±.015	83	1/4	13-1/2	14	.275 ±.006	13.475 ±.030
39	3/16	2-7/8	3-1/4	.210 ±.005	2.850 ±.015	84	1/4	14	14-1/2	.275 ±.006	13.975 ±.030
40	3/16	3	3-3/8	.210 ±.005	2.975 ±.015	85	1/4	14-1/2	15	.275 ±.006	14.475 ±.030
41	3/16	3-1/8	3-1/2	.210 ±.005	3.100 ±.015	86	1/4	15	15-1/2	.275 ±.006	14.975 ±.030
42	3/16	3-1/4	3-5/8	.210 ±.005	3.225 ±.015	87	1/4	15-1/2	16	.275 ±.006	15.475 ±.030
43	3/16	3-3/8	3-3/4	.210 ±.005	3.350 ±.015	88	1/4	4-1/2	5	.275 ±.006	4.475 ±.015
44	3/16	3-1/2	3-7/8	.210 ±.005	3.475 ±.015						
45	3/16	3-5/8	4	.210 ±.005	3.600 ±.015						

EXAMPLE OF PART NO.: AN6227-32 - PACKING WITH NOMINAL OD OF 2-3/8, ID OF 2 INCHES, MATERIAL - CLASS B OF SPEC MIL-P-5516.

(8) PACKING: ONE SET OF COLORED IDENTIFICATION DOTS IN ACCORDANCE WITH ANA BULLETIN NO. 419.

SEE SPECIFICATION MIL-P-5514 FOR INSTALLATION DETAILS.

DIMENSIONS IN INCHES.

FOR ACCEPTABLE PRODUCTS SEE QPL-5516.

ASSEMBLY DRAWINGS SHALL SPECIFY PACKINGS ONLY BY THEIR AN6227 NUMBER WITHOUT REFERENCE TO TYPE B, BUT REGARDLESS OF AN6227 CALLOUT, CLASS B PACKINGS ARE TO BE USED. DESIGNATION AN6227H IS TO BE USED FOR PACKAGING OR STOCKING.

(8) CERTAIN PROVISIONS OF THIS STANDARD ARE THE SUBJECT OF INTERNATIONAL STANDARDIZATION AGREEMENT ASCC AIR STD 17/27 AND STANAG 3444. WHEN REVISION OR CANCELLATION OF THIS STANDARD IS PROPOSED, THE DEPARTMENTAL CUSTODIANS WILL INFORM THEIR RESPECTIVE DEPARTMENTAL STANDARDIZATION OFFICES SO THAT APPROPRIATE ACTION MAY BE TAKEN RESPECTING THE INTERNATIONAL AGREEMENT CONCERNED.

PROCUREMENT SPECIFICATION	AIR FORCE-NAVY AERONAUTICAL STANDARD	AN6227
MIL-P-5516	PACKING, "O" RING HYDRAULIC	

Table 3. Military Standard, Preformed Packings (MS 28775)

SECTION A-A

PART NUMBER	ID IN.		ID (mm)		T IN.		T (mm)		APPROX MASS	
	MIN	MAX	MIN	MAX	MIN	MAX	MIN	MAX	LB/100	Kg/100
MS28775-001	0.025	0.033	0.635	0.838	0.037	0.043	0.940	1.092	.001	.0004
MS28775-002	0.038	0.046	0.965	1.168	0.047	0.063	1.194	1.346	.003	.0014
MS28775-003	0.052	0.060	1.321	1.524	0.057	0.063	1.448	1.600	.006	.0023
MS28775-004	0.065	0.075	1.65	1.90	0.067	0.073	1.702	1.854	.008	.0036
MS28775-005	0.096	0.106	2.44	2.69	0.067	0.073	1.702	1.854	.010	.0045
MS28775-006	0.109	0.119	2.77	3.02	0.067	0.073	1.702	1.854	.010	.0045
MS28775-007	0.140	0.150	3.56	3.81	0.067	0.073	1.702	1.854	.010	.0045
MS28775-008	0.171	0.181	4.34	4.60	0.067	0.073	1.702	1.854	.012	.0054
MS28775-009	0.203	0.213	5.16	5.41	0.067	0.073	1.702	1.854	.014	.0064
MS28775-010	0.234	0.244	5.94	6.20	0.067	0.073	1.702	1.854	.016	.0073
MS28775-011	0.296	0.306	7.52	7.77	0.067	0.073	1.702	1.854	.021	.0096
MS28775-012	0.359	0.369	9.12	9.37	0.067	0.073	1.702	1.854	.024	.011
MS28775-013	0.421	0.431	10.69	10.95	0.067	0.073	1.702	1.854	.028	.013
MS28775-014	0.484	0.494	12.29	12.55	0.067	0.073	1.702	1.854	.031	.014
MS28775-015	0.546	0.556	13.87	14.12	0.067	0.073	1.702	1.854	.034	.015
MS28775-016	0.609	0.619	15.47	15.72	0.067	0.073	1.702	1.854	.038	.017
MS28775-017	0.671	0.681	17.04	17.30	0.067	0.073	1.702	1.854	.041	.019
MS28775-018	0.734	0.744	18.64	18.90	0.067	0.073	1.702	1.854	.045	.020
MS28775-019	0.795	0.807	20.19	20.50	0.067	0.073	1.702	1.854	.048	.022
MS28775-020	0.858	0.870	21.79	22.10	0.067	0.073	1.702	1.854	.052	.024
MS28775-021	0.920	0.932	23.37	23.67	0.067	0.073	1.702	1.854	.055	.025
MS28775-022	0.983	0.995	24.97	25.27	0.067	0.073	1.702	1.854	.059	.027
MS28775-023	1.045	1.057	26.54	26.85	0.067	0.073	1.702	1.854	.062	.028
MS28775-024	1.108	1.120	28.14	28.45	0.067	0.073	1.702	1.854	.066	.030
MS28775-025	1.170	1.182	29.72	30.02	0.067	0.073	1.702	1.854	.069	.031
MS28775-026	1.233	1.245	31.32	31.62	0.067	0.073	1.702	1.854	.073	.033
MS28775-027	1.295	1.307	32.89	33.20	0.067	0.073	1.702	1.854	.076	.034
MS28775-028	1.358	1.370	34.49	34.80	0.067	0.073	1.702	1.854	.080	.036
MS28775-029	1.479	1.499	37.57	38.07	0.067	0.073	1.702	1.854	.086	.039
MS28775-030	1.604	1.624	40.74	41.25	0.067	0.073	1.702	1.854	.094	.043
MS28775-031	1.729	1.749	43.92	44.42	0.067	0.073	1.702	1.854	.101	.046
MS28775-032	1.854	1.874	47.09	47.60	0.067	0.073	1.702	1.854	.108	.049
MS28775-033	1.979	1.999	50.27	50.77	0.067	0.073	1.702	1.854	.114	.052
MS28775-034	2.104	2.124	53.44	53.95	0.067	0.073	1.702	1.854	.121	.055
MS28775-035	2.229	2.249	56.62	57.12	0.067	0.073	1.702	1.854	.128	.058
MS28775-036	2.354	2.374	59.79	60.30	0.067	0.073	1.702	1.854	.135	.061
MS28775-037	2.479	2.499	62.97	63.47	0.067	0.073	1.702	1.854	.142	.064
MS28775-038	2.604	2.624	66.14	66.65	0.067	0.073	1.702	1.854	.149	.068
MS28775-039	2.724	2.754	69.19	69.95	0.067	0.073	1.702	1.854	.156	.071
MS28775-040	2.849	2.879	72.36	73.13	0.067	0.073	1.702	1.854	.163	.074
MS28775-041	2.974	3.004	75.54	76.30	0.067	0.073	1.702	1.854	.170	.077
MS28775-042	3.224	3.254	81.89	82.65	0.067	0.073	1.702	1.854	.184	.083
MS28775-043	3.474	3.504	88.24	89.00	0.067	0.073	1.702	1.854	.198	.090
MS28775-044	3.724	3.754	94.59	95.35	0.067	0.073	1.702	1.854	.212	.096
MS28775-045	3.974	4.004	100.94	101.70	0.067	0.073	1.702	1.854	.226	.102
MS28775-046	4.224	4.254	107.29	108.06	0.067	0.073	1.702	1.854	.240	.109
MS28775-047	4.474	4.504	113.64	114.40	0.067	0.073	1.702	1.854	.254	.115
MS28775-048	4.724	4.754	119.99	120.75	0.067	0.073	1.702	1.854	.267	.121
MS28775-049	4.966	5.012	126.14	127.30	0.067	0.073	1.702	1.854	.281	.127
MS28775-050	5.216	5.262	132.49	133.65	0.067	0.073	1.702	1.854	.296	.134
MS28775-102	0.044	0.054	1.12	1.37	0.100	0.106	2.540	2.692	.018	.008
MS28775-103	0.076	0.086	1.93	2.18	0.100	0.106	2.540	2.692	.022	.010
MS28775-104	0.107	0.117	2.72	2.97	0.100	0.106	2.540	2.692	.026	.012
MS28775-105	0.138	0.148	3.51	3.76	0.100	0.106	2.540	2.692	.029	.013
MS28775-106	0.169	0.179	4.29	4.55	0.100	0.106	2.540	2.692	.034	.015
MS28775-107	0.201	0.211	5.11	5.36	0.100	0.106	2.540	2.692	.037	.017
MS28775-108	0.232	0.242	5.89	6.15	0.100	0.106	2.540	2.692	.041	.019
MS28775-109	0.294	0.304	7.47	7.72	0.100	0.106	2.540	2.692	.048	.022
MS28775-110	0.357	0.367	9.07	9.32	0.100	0.106	2.540	2.692	.056	.025

(D) ENTIRE STANDARD REVISED

User activities:
ARMY — EL
DSA — CS

Review activities:
ARMY — MU, MI, WC, AT
USAF — 82, 11
DSA — IS

1 SEP 75
(D) 1 SEP 75
(D) 22 AUG 68
(C) 6 JUL 65
(B) 23 JUN 59
REVISED (A) 12 JUL 57
APPROVED 12 JUL 57

P.A. 82	INTERNATIONAL INTEREST	TITLE	MILITARY STANDARD
Other Cust ARMY — AV NAVY — AS	ASCC 17/27	PACKING, PREFORMED, HYDRAULIC, +275°F, ("O" RING)	MS 28775
PROCUREMENT SPECIFICATION MIL-P-25732	SUPERSEDES: MS28784		SHEET 1 OF 6

DD, FORM 672-1 (Coordinated)   PREVIOUS EDITIONS OF THIS FORM ARE OBSOLETE.

Table 3. (continued)

PART NUMBER	ID IN.		ID (mm)		T IN.		T (mm)		APPROX MASS	
	MIN	MAX	MIN	MAX	MIN	MAX	MIN	MAX	LB/100	Kg/100
MS28775-111	0.419	0.429	10.64	10.90	0.100	0.106	2.540	2.692	.063	.029
MS28775-112	0.482	0.492	12.24	12.50	0.100	0.106	2.540	2.692	.071	.032
MS28775-113	0.544	0.554	13.82	14.07	0.100	0.106	2.540	2.692	.079	.036
MS28775-114	0.607	0.617	15.42	15.67	0.100	0.106	2.540	2.692	.086	.039
MS28775-115	0.669	0.679	16.99	17.25	0.100	0.106	2.540	2.692	.093	.042
MS28775-116	0.732	0.742	18.59	18.85	0.100	0.106	2.540	2.692	.101	.046
MS28775-117	0.793	0.805	20.14	20.45	0.100	0.106	2.540	2.692	.108	.049
MS28775-118	0.856	0.868	21.74	22.05	0.100	0.106	2.540	2.692	.116	.053
MS28775-119	0.918	0.930	23.32	23.62	0.100	0.106	2.540	2.692	.124	.056
MS28775-120	0.981	0.993	24.92	25.22	0.100	0.106	2.540	2.692	.131	.059
MS28775-121	1.043	1.055	26.49	26.80	0.100	0.106	2.540	2.692	.139	.063
MS28775-122	1.106	1.118	28.09	28.40	0.100	0.106	2.540	2.692	.146	.066
MS28775-123	1.168	1.180	29.67	29.97	0.100	0.106	2.540	2.692	.154	.070
MS28775-124	1.231	1.243	31.27	31.57	0.100	0.106	2.540	2.692	.161	.073
MS28775-125	1.293	1.305	32.84	33.15	0.100	0.106	2.540	2.692	.169	.077
MS28775-126	1.356	1.368	34.44	34.75	0.100	0.106	2.540	2.692	.176	.080
MS28775-127	1.418	1.430	36.02	36.32	0.100	0.106	2.540	2.692	.184	.083
MS28775-128	1.481	1.493	37.62	37.92	0.100	0.106	2.540	2.692	.191	.087
MS28775-129	1.539	1.559	39.09	39.60	0.100	0.106	2.540	2.692	.199	.090
MS28775-130	1.602	1.622	40.69	41.20	0.100	0.106	2.540	2.692	.207	.094
MS28775-131	1.664	1.684	42.27	42.77	0.100	0.106	2.540	2.692	.214	.097
MS28775-132	1.727	1.747	43.87	44.37	0.100	0.106	2.540	2.692	.222	.101
MS28775-133	1.789	1.809	45.44	45.95	0.100	0.106	2.540	2.692	.230	.104
MS28775-134	1.852	1.872	47.04	47.55	0.100	0.106	2.540	2.692	.236	.107
MS28775-135	1.915	1.935	48.64	49.15	0.100	0.106	2.540	2.692	.244	.111
MS28775-136	1.977	1.997	50.22	50.72	0.100	0.106	2.540	2.692	.252	.114
MS28775-137	2.040	2.060	51.82	52.32	0.100	0.106	2.540	2.692	.259	.117
MS28775-138	2.102	2.122	53.39	53.90	0.100	0.106	2.540	2.692	.267	.121
MS28775-139	2.165	2.185	54.99	55.50	0.100	0.106	2.540	2.692	.274	.124
MS28775-140	2.227	2.247	56.57	57.07	0.100	0.106	2.540	2.692	.282	.128
MS28775-141	2.290	2.310	58.17	58.67	0.100	0.106	2.540	2.692	.289	.131
MS28775-142	2.352	2.372	59.74	60.25	0.100	0.106	2.540	2.692	.297	.135
MS28775-143	2.415	2.435	61.34	61.85	0.100	0.106	2.540	2.692	.304	.138
MS28775-144	2.477	2.497	62.92	63.42	0.100	0.106	2.540	2.692	.312	.142
MS28775-145	2.540	2.560	64.52	65.02	0.100	0.106	2.540	2.692	.319	.145
MS28775-146	2.602	2.622	66.09	66.60	0.100	0.106	2.540	2.692	.327	.148
MS28775-147	2.660	2.690	67.56	68.33	0.100	0.106	2.540	2.692	.334	.152
MS28775-148	2.722	2.752	69.14	69.90	0.100	0.106	2.540	2.692	.342	.155
MS28775-149	2.785	2.815	70.74	71.50	0.100	0.106	2.540	2.692	.350	.159
MS28775-150	2.847	2.877	72.31	73.08	0.100	0.106	2.540	2.692	.357	.162
MS28775-151	2.972	3.002	75.49	76.25	0.100	0.106	2.540	2.692	.372	.169
MS28775-152	3.222	3.252	81.84	82.60	0.100	0.106	2.540	2.692	.402	.181
MS28775-153	3.472	3.502	88.19	88.95	0.100	0.106	2.540	2.692	.462	.210
MS28775-154	3.722	3.752	94.54	95.30	0.100	0.106	2.540	2.692	.462	.210
MS28775-155	3.972	4.002	100.89	101.65	0.100	0.106	2.540	2.692	.493	.224
MS28775-156	4.222	4.252	107.24	108.00	0.100	0.106	2.540	2.692	.523	.237
MS28775-157	4.472	4.502	113.59	114.35	0.100	0.106	2.540	2.692	.553	.251
MS28775-158	4.722	4.752	119.94	120.70	0.100	0.106	2.540	2.692	.583	.264
MS28775-159	4.972	5.002	126.29	127.06	0.100	0.106	2.540	2.692	.613	.278
MS28775-160	5.214	5.260	132.44	133.60	0.100	0.106	2.540	2.692	.643	.292
MS28775-161	5.464	5.510	138.79	139.96	0.100	0.106	2.540	2.692	.673	.306
MS28775-162	5.714	5.760	145.14	146.30	0.100	0.106	2.540	2.692	.703	.319
MS28775-163	5.964	6.010	151.49	152.65	0.100	0.106	2.540	2.692	.733	.332
MS28775-164	6.214	6.260	157.84	159.00	0.100	0.106	2.540	2.692	.763	.346
MS28775-165	6.464	6.510	164.19	165.35	0.100	0.106	2.540	2.692	.793	.360
MS28775-166	6.714	6.760	170.54	171.70	0.100	0.106	2.540	2.692	.823	.373
MS28775-167	6.964	7.010	176.89	178.05	0.100	0.106	2.540	2.692	.853	.387
MS28775-168	7.207	7.267	183.06	184.58	0.100	0.106	2.540	2.692	.883	.401
MS28775-169	7.457	7.517	189.41	190.93	0.100	0.106	2.540	2.692	.913	.414
MS28775-170	7.707	7.767	195.76	197.28	0.100	0.106	2.540	2.692	.943	.428
MS28775-171	7.957	8.017	202.11	203.63	0.100	0.106	2.540	2.692	.973	.441
MS28775-172	8.207	8.267	208.46	209.98	0.100	0.106	2.540	2.692	1.003	.455
MS28775-173	8.457	8.517	214.81	216.33	0.100	0.106	2.540	2.692	1.033	.469
MS28775-174	8.707	8.767	221.16	222.68	0.100	0.106	2.540	2.692	1.064	.483
MS28775-175	8.967	9.017	227.51	229.03	0.100	0.106	2.540	2.692	1.094	.496
MS28775-176	9.207	9.267	233.86	235.38	0.100	0.106	2.540	2.692	1.124	.510
MS28775-177	9.457	9.517	240.21	241.73	0.100	0.106	2.540	2.692	1.154	.523
MS28775-178	9.707	9.767	246.56	248.08	0.100	0.106	2.540	2.692	1.184	.537

User activities:
ARMY — LL
DSA — CS

Review activities:
ARMY — MU, MI, WC, AT
USAF — 82, 11
DSA — 13

This military standard is approved for use by all Departments and Agencies of the Department of Defense. Selection for all new engineering and design application and for repetitive use shall be made from this document.

(D) FOR CHANGES SEE SHEET 1 THRU 6

REVISED (D)

APPROVED 12 JUL 57

(D) ENTIRE STANDARD REVISED

P.A. 82	INTERNATIONAL INTEREST	TITLE	MILITARY STANDARD
Other Cust ARMY–AV NAVY–AS	ASCC 17/27 (D)	PACKING, PREFORMED, HYDRAULIC, +275°F, ("O" RING)	MS 28775
PROCUREMENT SPECIFICATION MIL-P-25732	SUPERSEDES: MS28784		SHEET 2 OF 8

DD 672-1 (Coordinated)

PREVIOUS EDITIONS OF THIS FORM ARE OBSOLETE.

10

Table 3. (continued)

PART NUMBER	ID IN. MIN	ID IN. MAX	ID (mm) MIN	ID (mm) MAX	T IN. MIN	T IN. MAX	T (mm) MIN	T (mm) MAX	APPROX MASS LB/100	APPROX MASS Kg/100
MS28775-201	0.166	0.176	4.22	4.47	0.135	0.143	3.429	3.632	.068	.031
MS28775-202	0.229	0.239	5.82	6.07	0.135	0.143	3.429	3.632	.082	.037
MS28775-203	0.291	0.301	7.39	7.65	0.135	0.143	3.429	3.632	.096	.043
MS28775-204	0.354	0.364	8.99	9.25	0.135	0.143	3.429	3.632	.109	.049
MS28775-205	0.416	0.426	10.57	10.82	0.135	0.143	3.429	3.632	.123	.056
MS28775-206	0.479	0.489	12.17	12.42	0.135	0.143	3.429	3.632	.137	.062
MS28775-207	0.541	0.551	13.74	14.00	0.135	0.143	3.429	3.632	.150	.068
MS28775-208	0.604	0.614	15.34	15.60	0.135	0.143	3.429	3.632	.164	.074
MS28775-209	0.666	0.676	16.92	17.17	0.135	0.143	3.429	3.632	.178	.081
MS28775-210	0.728	0.740	18.49	18.80	0.135	0.143	3.429	3.632	.191	.087
MS28775-211	0.790	0.802	20.07	20.37	0.135	0.143	3.429	3.632	.205	.093
MS28775-212	0.853	0.865	21.67	21.97	0.135	0.143	3.429	3.632	.219	.099
MS28775-213	0.915	0.927	23.24	23.55	0.135	0.143	3.429	3.632	.232	.105
MS28775-214	0.978	0.990	24.84	25.15	0.135	0.143	3.429	3.632	.246	.112
MS28775-215	1.040	1.052	26.42	26.72	0.135	0.143	3.429	3.632	.260	.118
MS28775-216	1.103	1.115	28.02	28.32	0.135	0.143	3.429	3.632	.274	.124
MS28775-217	1.165	1.177	29.59	29.90	0.135	0.143	3.429	3.632	.287	.130
MS28775-218	1.228	1.240	31.19	31.50	0.135	0.143	3.429	3.632	.302	.137
MS28775-219	1.290	1.302	32.77	33.07	0.135	0.143	3.429	3.632	.315	.143
MS28775-220	1.353	1.365	34.37	34.67	0.135	0.143	3.429	3.632	.328	.149
MS28775-221	1.415	1.427	35.94	36.25	0.135	0.143	3.429	3.632	.342	.155
MS28775-222	1.478	1.490	37.54	37.85	0.135	0.143	3.429	3.632	.356	.161
MS28775-223	1.599	1.619	40.61	41.12	0.135	0.143	3.429	3.632	.383	.174
MS28775-224	1.724	1.744	43.79	44.30	0.135	0.143	3.429	3.632	.411	.186
MS28775-225	1.849	1.869	46.96	47.47	0.135	0.143	3.429	3.632	.438	.199
MS28775-226	1.974	1.994	50.14	50.65	0.135	0.143	3.429	3.632	.466	.211
MS28775-227	2.099	2.119	53.31	53.82	0.135	0.143	3.429	3.632	.493	.224
MS28775-228	2.224	2.244	56.49	57.00	0.135	0.143	3.429	3.632	.529	.240
MS28775-229	2.349	2.369	59.66	60.17	0.135	0.143	3.429	3.632	.548	.249
MS28775-230	2.474	2.494	62.84	63.35	0.135	0.143	3.429	3.632	.575	.261
MS28775-231	2.599	2.619	66.01	66.52	0.135	0.143	3.429	3.632	.603	.274
MS28775-232	2.719	2.749	69.06	69.82	0.135	0.143	3.429	3.632	.630	.286
MS28775-233	2.844	2.874	72.24	73.00	0.135	0.143	3.429	3.632	.657	.298
MS28775-234	2.969	2.999	75.41	76.17	0.135	0.143	3.429	3.632	.685	.311
MS28775-235	3.094	3.124	78.59	79.35	0.135	0.143	3.429	3.632	.712	.323
MS28775-236	3.219	3.249	81.76	82.52	0.135	0.143	3.429	3.632	.740	.336
MS28775-237	3.344	3.374	84.94	85.70	0.135	0.143	3.429	3.632	.767	.348
MS28775-238	3.469	3.499	88.11	88.87	0.135	0.143	3.429	3.632	.794	.360
MS28775-239	3.594	3.624	91.29	92.05	0.135	0.143	3.429	3.632	.822	.373
MS28775-240	3.719	3.749	94.46	95.22	0.135	0.143	3.429	3.632	.849	.385
MS28775-241	3.844	3.874	97.64	98.40	0.135	0.143	3.429	3.632	.877	.398
MS28775-242	3.969	3.999	100.81	101.57	0.135	0.143	3.429	3.632	.904	.410
MS28775-243	4.094	4.124	103.99	104.75	0.135	0.143	3.429	3.632	.932	.423
MS28775-244	4.219	4.249	107.16	107.92	0.135	0.143	3.429	3.632	.959	.435
MS28775-245	4.344	4.374	110.34	111.10	0.135	0.143	3.429	3.632	.986	.447
MS28775-246	4.469	4.499	113.51	114.27	0.135	0.143	3.429	3.632	1.014	.460
MS28775-247	4.594	4.624	116.69	117.45	0.135	0.143	3.429	3.632	1.041	.472
MS28775-248	4.719	4.749	119.86	120.62	0.135	0.143	3.429	3.632	1.068	.484
MS28775-249	4.844	4.874	123.04	123.80	0.135	0.143	3.429	3.632	1.096	.497
MS28775-250	4.969	4.999	126.21	126.97	0.135	0.143	3.429	3.632	1.123	.509
MS28775-251	5.086	5.132	129.18	130.35	0.135	0.143	3.429	3.632	1.151	.522
MS28775-252	5.211	5.257	132.36	133.53	0.135	0.143	3.429	3.632	1.178	.534
MS28775-253	5.336	5.382	135.53	136.70	0.135	0.143	3.429	3.632	1.206	.547
MS28775-254	5.461	5.507	138.71	139.88	0.155	0.143	3.429	3.632	1.233	.559
MS28775-255	5.586	5.632	141.88	143.05	0.135	0.143	3.429	3.632	1.260	.572
MS28775-256	5.711	5.757	145.06	146.23	0.135	0.143	3.429	3.632	1.288	.584
MS28775-257	5.836	5.882	148.23	149.40	0.135	0.143	3.429	3.632	1.315	.596
MS28775-258	5.961	6.007	151.41	152.58	0.135	0.143	3.429	3.632	1.343	.609
MS28775-259	6.211	6.257	157.76	158.93	0.135	0.143	3.429	3.632	1.397	.634
MS28775-260	6.461	6.507	164.11	165.28	0.135	0.143	3.429	3.632	1.452	.659
MS28775-261	6.711	6.757	170.46	171.63	0.135	0.143	3.429	3.632	1.507	.684
MS28775-262	6.961	7.007	176.81	177.98	0.135	0.143	3.429	3.632	1.562	.708
MS28775-263	7.204	7.264	182.98	184.51	0.135	0.143	3.429	3.632	1.617	.733
MS28775-264	7.454	7.514	189.33	190.86	0.135	0.143	3.429	3.632	1.672	.758
MS28775-265	7.704	7.764	195.68	197.21	0.135	0.143	3.429	3.632	1.726	.783
MS28775-266	7.954	8.014	202.03	203.56	0.135	0.143	3.429	3.632	1.781	.808
MS28775-267	8.204	8.264	208.38	209.91	0.135	0.143	3.429	3.632	1.836	.833
MS28775-268	8.454	8.514	214.73	216.26	0.135	0.143	3.429	3.632	1.891	.858
MS28775-269	8.704	8.764	221.08	222.61	0.135	0.143	3.429	3.632	1.946	.883
MS28775-270	8.954	9.014	227.43	228.96	0.135	0.143	3.429	3.632	2.000	.907

(D) ENTIRE STANDARD REVISED

P.A. 32	INTERNATIONAL INTEREST	TITLE		
Other Cust	ASCC 17/27	PACKING, PREFORMED, HYDRAULIC, +275°F, ("O" RING)	**MILITARY STANDARD**	
ARMY – AV NAVY – AS	(D)		**MS 28775**	
PROCUREMENT SPECIFICATION MIL-P-25732	SUPERSEDES: MS29744		SHEET 3	OF 6

DD, FORM 872-1 (Coordinated)

PREVIOUS EDITIONS OF THIS FORM ARE OBSOLETE.

APPROVED 12 JUL 57   REVISED (D) FOR CHANGES SEE SHEET 1 THRU 6

Review activities:
ARMY — MU, MI, WC, AT
USAF — B2, 11
DSA — IS

User activities:
ARMY — EL
DSA — CS

This military standard is approved for use by all Departments and Agencies of the Department of Defense. Selection for all new engineering and design application and for repetitive use shall be made from this document.

Table 3. (continued)

FED. SUP CLASS
5330

PART NUMBER	ID IN.		ID (mm)		T IN.		T (mm)		APPROX MASS	
	MIN	MAX	MIN	MAX	MIN	MAX	MIN	MAX	LB/100	Kg/100
MS28775-271	9.204	9.264	233.78	235.71	0.135	0.143	3.429	3.632	2.056	.932
MS28775-272	9.454	9.514	240.13	241.66	0.135	0.143	3.429	3.632	2.110	.957
MS28775-273	9.704	9.784	246.48	248.01	0.135	0.143	3.429	3.632	2.165	.982
MS28775-274	9.954	10.014	252.83	254.36	0.135	0.143	3.429	3.632	2.220	1.007
MS28775-275	10.454	10.514	265.63	267.06	0.135	0.143	3.429	3.632	2.329	1.056
MS28775-276	10.954	11.014	278.23	279.76	0.135	0.143	3.429	3.632	2.439	1.106
MS28775-277	11.454	11.514	290.93	292.46	0.135	0.143	3.429	3.632	2.549	1.156
MS28775-278	11.954	12.014	303.63	305.16	0.135	0.143	3.429	3.632	2.658	1.206
MS28775-279	12.954	13.014	329.03	330.56	0.135	0.143	3.429	3.632	2.878	1.305
MS28775-280	13.954	14.014	354.43	355.96	0.135	0.143	3.429	3.632	3.097	1.405
MS28775-281	14.954	15.014	379.83	381.36	0.135	0.143	3.429	3.632	3.317	1.505
MS28775-282	15.910	16.000	404.11	406.40	0.135	0.143	3.429	3.632	3.531	1.602
MS28775-283	16.910	17.000	429.51	431.80	0.135	0.143	3.429	3.632	3.751	1.701
MS28775-284	17.910	18.000	454.91	457.20	0.135	0.143	3.429	3.632	3.968	1.800
MS28775-309	0.407	0.417	10.34	10.59	0.205	0.215	5.21	5.46	.311	.141
MS28775-310	0.470	0.480	11.94	12.19	0.205	0.215	5.21	5.46	.343	.155
MS28775-311	0.532	0.542	13.51	13.77	0.205	0.215	5.21	5.46	.374	.170
MS28775-312	0.596	0.606	15.11	15.37	0.205	0.215	5.21	5.46	.405	.184
MS28775-313	0.657	0.667	16.69	16.94	0.205	0.215	5.21	5.46	.436	.198
MS28775-314	0.719	0.731	18.26	18.57	0.206	0.215	5.21	5.46	.468	.212
MS28775-315	0.781	0.793	19.84	20.14	0.206	0.215	5.21	5.46	.499	.226
MS28775-316	0.844	0.856	21.44	21.74	0.206	0.215	5.21	5.46	.530	.241
MS28775-317	0.908	0.918	23.01	23.32	0.206	0.215	5.21	5.46	.562	.255
MS28775-318	0.969	0.981	24.61	24.92	0.206	0.215	5.21	5.46	.593	.269
MS28775-319	1.031	1.043	26.19	26.49	0.205	0.215	5.21	5.46	.624	.283
MS28775-320	1.094	1.106	27.79	28.09	0.205	0.215	5.21	5.46	.656	.297
MS28775-321	1.156	1.168	29.36	29.57	0.205	0.215	5.21	5.46	.687	.312
MS28775-322	1.219	1.231	30.96	31.27	0.206	0.215	5.21	5.46	.718	.326
MS28775-323	1.281	1.293	32.54	32.84	0.205	0.215	5.21	5.46	.749	.340
MS28775-324	1.344	1.356	34.14	34.44	0.206	0.215	5.21	5.46	.781	.354
MS28775-325	1.466	1.486	37.21	37.72	0.205	0.215	5.21	5.46	.843	.382
MS28775-326	1.590	1.610	40.39	40.89	0.205	0.215	5.21	5.46	.906	.411
MS28775-327	1.715	1.735	43.56	44.07	0.206	0.215	5.21	5.46	.969	.440
MS28775-328	1.840	1.860	46.74	47.24	0.205	0.215	5.21	5.46	1.031	.468
MS28775-329	1.965	1.985	49.91	50.42	0.205	0.215	5.21	5.46	1.094	.496
MS28775-330	2.090	2.110	53.09	53.59	0.205	0.215	5.21	5.46	1.156	.524
MS28775-331	2.215	2.235	56.26	56.77	0.205	0.215	5.21	5.46	1.219	.553
MS28775-332	2.340	2.360	59.44	59.94	0.205	0.215	5.21	5.46	1.282	.582
MS28775-333	2.465	2.485	62.61	63.12	0.205	0.215	5.21	5.46	1.344	.610
MS28775-334	2.590	2.610	65.79	66.29	0.206	0.215	5.21	5.45	1.407	.638
MS28775-335	2.710	2.740	68.83	69.60	0.205	0.215	5.21	5.46	1.469	.666
MS28775-336	2.835	2.865	72.01	72.77	0.205	0.215	5.21	5.46	1.532	.695
MS28775-337	2.960	2.990	75.18	75.95	0.205	0.215	5.21	5.46	1.594	.723
MS28775-338	3.085	3.115	78.36	79.12	0.205	0.215	5.21	5.46	1.657	.752
MS28775-339	3.210	3.240	81.53	82.30	0.205	0.215	5.21	5.46	1.719	.780
MS28775-340	3.335	3.365	84.71	85.47	0.205	0.215	5.21	5.46	1.782	.808
MS28775-341	3.460	3.490	87.88	88.65	0.206	0.215	5.21	5.46	1.845	.837
MS28775-342	3.586	3.615	91.06	91.82	0.205	0.215	5.21	5.46	1.907	.865
MS28775-343	3.710	3.740	94.23	95.00	0.205	0.215	5.21	5.46	1.970	.894
MS28775-344	3.836	3.865	97.41	98.17	0.205	0.215	5.21	5.46	2.032	.922
MS28775-345	3.960	3.990	100.58	101.35	0.205	0.215	5.21	5.46	2.095	.950
MS28775-346	4.086	4.115	103.76	104.52	0.205	0.215	5.21	5.46	2.157	.978
MS28775-347	4.210	4.240	106.93	107.70	0.205	0.215	5.21	5.46	2.220	1.007
MS28775-348	4.335	4.365	110.11	110.87	0.205	0.215	5.21	5.46	2.282	1.035
MS28775-349	4.460	4.490	113.28	114.05	0.205	0.215	5.21	5.46	2.345	1.064
MS28775-350	4.586	4.615	116.46	117.22	0.205	0.215	5.21	5.46	2.412	1.094
MS28775-351	4.710	4.740	119.63	120.40	0.206	0.215	5.21	5.46	2.474	1.122
MS28775-352	4.836	4.865	122.81	123.57	0.205	0.215	5.21	5.46	2.537	1.151
MS28775-353	4.960	4.990	125.98	126.75	0.206	0.215	5.21	5.46	2.600	1.179
MS28775-354	5.077	5.123	128.96	130.12	0.205	0.215	5.21	5.46	2.662	1.207
MS28775-356	5.202	5.248	132.13	133.30	0.206	0.215	5.21	5.46	2.725	1.236
MS28775-356	5.327	5.373	135.31	136.47	0.206	0.215	5.21	5.46	2.788	1.265
MS28775-357	5.452	5.498	138.48	139.56	0.206	0.215	5.21	5.46	2.851	1.293
MS28775-358	5.577	5.623	141.66	142.82	0.206	0.215	5.21	5.46	2.913	1.321
MS28775-359	5.702	5.748	144.83	146.00	0.206	0.215	5.21	5.46	2.976	1.350
MS28775-360	5.827	5.873	148.01	149.17	0.206	0.215	5.21	5.46	3.038	1.378
MS28775-361	5.952	5.998	151.18	152.35	0.206	0.215	5.21	5.46	3.101	1.407
MS28775-362	6.202	6.248	157.53	158.70	0.206	0.215	5.21	5.46	3.226	1.463
MS28775-363	6.452	6.498	163.88	165.06	0.206	0.215	5.21	5.46	3.352	1.520
MS28775-364	6.702	6.748	170.23	171.40	0.206	0.215	5.21	5.46	3.477	1.577
MS28775-365	6.952	6.998	176.58	177.75	0.206	0.215	5.21	5.46	3.603	1.634

Review activities:
ARMY — MU, MI, WC, AT
USAF — 82, 11
DSA — IS

User activities:
ARMY — EL
DSA — CS

APPROVED 12 JUL 57 REVISED (D) FOR CHANGES SEE SHEET I THRU 6

(D) ENTIRE STANDARD REVISED

P.A. 82	INTERNATIONAL INTEREST	TITLE	MILITARY STANDARD
Other Cust	ASCC 17/27	PACKING, PREFORMED, HYDRAULIC, +275°F, ("O" RING)	MS 28775
ARMY – AV			
NAVY – AS	(D)		
PROCUREMENT SPECIFICATION MIL-P-25732	SUPERSEDES: MS28784		SHEET 4 OF 6

DD ꞏꞏꞏ 672-1 (Coordinated)   PREVIOUS EDITIONS OF THIS FORM ARE OBSOLETE.

The military standard is approved for use by all Departments and Agencies of the Department of Defense. Retention for all users unregarding and design application and for the equivalent use shall be made from this document.

Table 3. (continued)

PART NUMBER	ID IN.		ID (mm)		T IN.		T (mm)		APPROX MASS	
	MIN	MAX	MIN	MAX	MIN	MAX	MIN	MAX	LB/100	Kg/100
MS28775-366	7.195	7.256	182.75	184.28	0.206	0.215	5.21	5.46	3.728	1.691
MS28775-367	7.445	7.506	189.10	190.65	0.206	0.215	5.21	5.46	3.864	1.748
MS28775-368	7.695	7.755	195.45	196.98	0.206	0.215	5.21	5.46	3.979	1.806
MS28775-369	7.945	8.006	201.80	203.33	0.206	0.215	5.21	5.46	4.104	1.862
MS28775-370	8.195	8.255	208.15	209.68	0.206	0.215	5.21	5.46	4.229	1.918
MS28775-371	8.445	8.506	214.50	216.03	0.206	0.215	5.21	5.46	4.355	1.975
MS28775-372	8.695	8.755	220.36	222.38	0.206	0.215	5.21	5.46	4.480	2.032
MS28775-373	8.945	9.005	227.20	228.73	0.206	0.215	5.21	5.46	4.605	2.089
MS28775-374	9.195	9.255	233.55	235.08	0.206	0.215	5.21	5.46	4.731	2.144
MS28775-375	9.445	9.506	239.90	241.43	0.206	0.215	5.21	5.46	4.856	2.203
MS28775-376	9.696	9.755	246.25	247.78	0.206	0.215	5.21	5.46	4.981	2.259
MS28775-377	9.945	10.006	252.60	254.13	0.206	0.215	5.21	5.46	5.107	2.317
MS28775-378	10.445	10.506	265.30	266.83	0.206	0.215	5.21	5.46	5.358	2.430
MS28775-379	10.945	11.006	273.00	279.53	0.205	0.215	5.21	5.46	5.608	2.544
MS28775-380	11.445	11.506	290.70	292.23	0.206	0.215	5.21	5.46	5.869	2.658
MS28775-381	11.945	12.006	303.40	304.93	0.205	0.215	5.21	5.46	6.110	2.771
MS28775-382	12.945	13.006	328.80	330.33	0.206	0.215	5.21	5.46	6.611	2.999
MS28775-383	13.945	14.006	354.20	355.73	0.205	0.215	5.21	5.46	7.113	3.226
MS28775-384	14.945	15.006	379.60	381.13	0.206	0.215	5.21	5.46	7.614	3.454
MS28775-385	15.910	16.000	404.11	406.40	0.206	0.215	5.21	5.46	8.106	3.676
MS28775-386	16.910	17.000	429.51	431.80	0.206	0.215	5.21	5.46	8.607	3.904
MS28775-387	17.910	18.000	454.91	457.20	0.206	0.215	5.21	5.46	9.106	4.131
MS28775-388	18.910	19.000	480.31	482.60	0.206	0.215	5.21	5.46	9.609	4.360
MS28775-389	19.910	20.000	505.71	508.00	0.206	0.215	5.21	5.46	10.111	4.586
MS28775-390	20.910	21.000	531.11	533.40	0.206	0.215	5.21	5.46	10.612	4.814
MS28775-391	21.910	22.000	556.51	558.80	0.206	0.215	5.21	5.46	11.114	5.041
MS28775-392	22.880	23.000	581.2	584.2	0.206	0.215	5.21	5.46	11.608	5.265
MS28775-393	23.880	24.000	606.6	609.6	0.205	0.215	5.21	5.46	12.109	5.493
MS28775-394	24.880	25.000	632.0	635.0	0.206	0.215	5.21	5.46	12.610	5.730
MS28775-395	25.880	26.000	657.4	660.4	0.206	0.215	5.21	5.46	13.112	5.948
MS28775-425	4.460	4.490	113.28	114.06	0.269	0.281	6.83	7.14	4.077	1.849
MS28775-426	4.585	4.615	116.46	117.22	0.269	0.281	6.83	7.14	4.185	1.898
MS28775-427	4.710	4.740	119.63	120.40	0.269	0.281	6.83	7.14	4.292	1.947
MS28775-428	4.835	4.865	122.81	123.57	0.269	0.281	6.83	7.14	4.399	1.995
MS28775-429	4.960	4.990	125.98	126.75	0.269	0.281	6.83	7.14	4.506	2.044
MS28775-430	5.077	5.123	128.96	130.12	0.269	0.281	6.83	7.14	4.614	2.093
MS28775-431	5.202	5.248	132.13	133.30	0.269	0.281	6.83	7.14	4.721	2.141
MS28775-432	5.327	5.373	135.31	136.47	0.269	0.281	6.83	7.14	4.828	2.190
MS28775-433	5.452	5.498	138.48	139.65	0.269	0.281	6.83	7.14	4.935	2.239
MS28775-434	5.577	5.623	141.66	142.82	0.269	0.281	6.83	7.14	5.042	2.287
MS28775-435	5.702	5.748	144.83	146.00	0.269	0.281	6.83	7.14	5.150	2.336
MS28775-436	5.827	5.873	148.01	149.17	0.269	0.281	6.83	7.14	5.257	2.386
MS28775-437	5.952	5.998	151.18	152.35	0.269	0.281	6.83	7.14	5.364	2.433
MS28775-438	6.202	6.248	157.53	158.70	0.269	0.281	6.83	7.14	5.579	2.531
MS28775-439	6.452	6.498	163.88	165.06	0.269	0.281	6.83	7.14	5.794	2.628
MS28775-440	6.702	6.748	170.23	171.40	0.269	0.281	6.83	7.14	6.008	2.725
MS28775-441	6.952	6.998	176.58	177.75	0.269	0.281	6.83	7.14	6.223	2.823
MS28775-442	7.195	7.255	182.75	184.28	0.269	0.281	6.83	7.14	6.438	2.920
MS28775-443	7.445	7.506	189.10	190.63	0.269	0.281	6.83	7.14	6.652	3.017
MS28775-444	7.695	7.755	195.45	196.98	0.269	0.281	6.83	7.14	6.867	3.115
MS28775-445	7.945	8.006	201.80	203.33	0.269	0.281	6.83	7.14	7.081	3.212
MS28775-446	8.445	8.506	214.50	216.03	0.269	0.281	6.83	7.14	7.510	3.407
MS28775-447	8.945	9.006	227.20	228.73	0.269	0.281	6.83	7.14	7.940	3.602
MS28775-448	9.445	9.506	239.90	241.43	0.269	0.281	6.83	7.14	8.369	3.796
MS28775-449	9.945	10.006	252.60	254.13	0.269	0.281	6.83	7.14	8.798	3.991
MS28775-450	10.445	10.506	265.30	266.83	0.269	0.281	6.83	7.14	9.227	4.186
MS28775-451	10.945	11.006	278.00	279.53	0.269	0.281	6.83	7.14	9.656	4.380
MS28775-452	11.445	11.506	290.70	292.23	0.269	0.281	6.83	7.14	10.086	4.575
MS28775-453	11.945	12.006	303.40	304.93	0.269	0.281	6.83	7.14	10.515	4.770
MS28775-454	12.445	12.506	316.10	317.63	0.269	0.281	6.83	7.14	10.944	4.964
MS28775-455	12.945	13.006	328.80	330.33	0.269	0.281	6.83	7.14	11.373	5.159
MS28775-456	13.445	13.506	341.50	343.03	0.269	0.281	6.83	7.14	11.802	5.353
MS28775-457	13.945	14.006	354.20	355.73	0.269	0.281	6.83	7.14	12.231	5.548
MS28775-458	14.445	14.506	366.90	368.43	0.269	0.281	6.83	7.14	12.660	5.742
MS28775-459	14.945	15.006	379.60	381.13	0.269	0.281	6.83	7.14	13.090	5.938
MS28775-460	15.445	15.506	392.30	393.83	0.269	0.281	6.83	7.14	13.519	6.132
MS28775-461	15.910	16.000	404.11	406.40	0.269	0.281	6.83	7.14	13.961	6.333
MS28775-462	16.410	16.500	416.81	419.10	0.269	0.281	6.83	7.14	14.391	6.528
MS28775-463	16.910	17.000	429.51	431.80	0.269	0.281	6.83	7.14	14.821	6.723
MS28775-464	17.410	17.500	442.21	444.50	0.269	0.281	6.83	7.14	15.251	6.918
MS28775-465	17.910	18.000	454.91	457.20	0.269	0.281	6.83	7.14	15.681	7.113

(D) ENTIRE STANDARD REVISED

P.A. 82	INTERNATIONAL INTEREST	TITLE	MILITARY STANDARD
Other Cust	ASCC 17/27	PACKING, PREFORMED, HYDRAULIC, + 275°F, ("O" RING)	MS 28775
ARMY — AV	(D)		
NAVY — AS			
PROCUREMENT SPECIFICATION	SUPERSEDES:		SHEET 5 OF 6
MIL-P-25732	MS28784		

DD, FORM 672-1 (Coordinated)   PREVIOUS EDITIONS OF THIS FORM ARE OBSOLETE.

User activities:
ARMY — EL
DSA — CS

Review activities:
ARMY — MU, MI, WC, AT
USAF — 82, 11
DSA — IS

This military standard is approved for use by all Departments and Agencies of the Department of Defense. Selection for all new engineering and design application and for repetitive use shall be made from this document.

APPROVED 12 JUL 57   REVISED (D) FOR CHANGES SEE SHEETS 1 THRU 6

13

Table 3. (continued)

PART NUMBER	ID IN.		ID (mm)		T IN.		T (mm)		APPROX MASS	
	MIN	MAX	MIN	MAX	MIN	MAX	MIN	MAX	LB/100	Kg/100
MS28775-466	18.410	18.500	467.61	469.90	0.269	0.281	6.83	7.14	16.112	7.307
MS28775-467	18.910	19.000	480.31	482.60	0.269	0.281	6.83	7.14	16.542	7.503
MS28775-468	19.410	19.500	493.01	495.30	0.269	0.281	6.83	7.14	16.972	7.698
MS28775-469	19.910	20.000	505.71	508.00	0.269	0.281	6.83	7.14	17.402	7.894
MS28775-470	20.910	21.000	531.11	533.40	0.269	0.281	6.83	7.14	18.262	8.284
MS28775-471	21.910	22.000	556.51	558.80	0.269	0.281	6.83	7.14	19.122	8.674
MS28775-472	22.880	23.000	581.2	584.2	0.269	0.281	6.83	7.14	19.970	9.058
MS28775-473	23.880	24.000	606.6	609.6	0.269	0.281	6.83	7.14	20.830	9.448
MS28775-474	24.880	25.000	632.0	635.0	0.269	0.281	6.83	7.14	21.690	9.839
MS28775-475	25.880	26.000	657.4	660.4	0.269	0.281	6.83	7.14	22.550	10.229

* O-RING SIZES -013 THROUGH -028, -117 THROUGH -149, AND -223 THROUGH -247 ARE INTENDED ONLY FOR USE AS STATIC SEALS, AND ARE NOT TO BE USED IN APPLICATIONS INVOLVING RECIPROCATING OR ROTARY INVOLVEMENT.

EXAMPLE OF PART NO. MS28775-211 - PACKING WITH ACTUAL ID DIA .796 IN., T DIA .139 IN., OR 20.22 MM ID DIA, 3.530 MM T DIA.

RINGS MAY BE OFF-REGISTER, DUE TO MOLD MISALIGNMENT, NOT TO EXCEED 0.003 INCH, PROVIDED ALL CROSS-SECTIONAL DIAMETERS, INCLUDING THE PROTRUSIONS (FLASH EXCLUDED) OF BOTH HALVES, WHERE APPLICABLE, SATISFY "T" DIMENSIONS.
CERTAIN PROVISION (DIMENSIONS FOR SIZES -001 THRU -050, -106 THRU -178, -210 THRU -281, -325 THRU -349, -425 THRU -460) OF THIS STANDARD ARE THE SUBJECT OF INTERNATIONAL STANDARDIZATION AGREEMENT ASCC AIR STD 17/27 AND NATO STANAG 3444. WHEN REVISION OR CANCELLATION OF THIS STANDARD IS PROPOSED WHICH WILL EFFECT OR VIOLATE THE INTERNATIONAL AGREEMENT CONCERNED, THE PREPARING ACTIVITY WILL TAKE APPROPRIATE RECONCILIATION ACTION THROUGH INTERNATIONAL STANDARDIZATION CHANNELS, INCLUDING DEPARTMENTAL STANDARDIZATION OFFICES, IF REQUIRED.

FOR DESIGN FEATURE PURPOSES, THIS STANDARD TAKES PRECEDENCE OVER PROCUREMENT DOCUMENTS REFERENCED HEREIN.
REFERENCED DOCUMENTS SHALL BE OF THE ISSUE IN EFFECT ON DATE OF INVITATIONS FOR BID.

User activities:
ARMY — EL
DSA — CS

Review activities:
ARMY — MU, MI, WC, AT
USAF — 11
DSA — 13

This military standard is approved for use by all Departments and Agencies of the Department of Defense. Selection for all new engineering and design application and for repetitive use shall be made from this document.

APPROVED 12 JUL 57   REVISED (D) FOR CHANGES SEE SHEETS 1 THRU 6

(D) ENTIRE STANDARD REVISED

P.A. 82	INTERNATIONAL INTEREST	TITLE	MILITARY STANDARD
Other Cust	ASCC 17/27	PACKING, PREFORMED, HYDRAULIC, +275°F, ("O" RING)	MS 28775
APMY - AV			
NAVY - AS (D)			

| PROCUREMENT SPECIFICATION NIL-P-25732 | SUPERSEDES: MS28784 | SHEET 6 OF 6 |

DD FORM 672-1 (Coordinated)   PREVIOUS EDITIONS OF THIS FORM ARE OBSOLETE.

Table 4. O-Rings for Tube Fittings

PARKER SERIES 3-XX O-RING SIZE CROSS-REFERENCE TABLE

These O-rings are intended for use with internal straight-thread fluid connection bosses and tube fittings. Ref. AND 10049, AND 10050, MS33656, MS33657, SAE straight-thread O-ring boss and mating swivel and adjustable style fittings.*

O-ring series	Compound specification	Parker compound no.
	Sequence A	
M25988/1[a]	MIL-R-25988 Cl. 1, Gr. 70[a]	L677-70[b]
M25988/3[a]	MIL-R-25988 Cl. 1, Gr. 60[a]	L737-65[b]
M25988/4[a]	MIL-R-25988 Cl. 1, Gr. 80[a]	L806-80[b]
M83248/1	MIL-R-83248 Cl. 1	V747-75
M83248/2	MIL-R-83248 Cl. 2	V709-90
	Sequence B	
MS9020	AMS7271	N506-65
MS9355	AMS7272	N287-70[b]
MS29512	MIL-P-5315	N602-70

Table 4. (continued)

1												
O-ring series			Compound specification				Parker compound no.					

Sequence C

O-ring series	Compound specification	Parker compound no.
AN6290(OBS)	MIL-P-5510	N507-90
MS28778	MIL-P-5510	N507-90
NAS617	MIL-R-7362	47-071
NAS1595	MIL-R-25897, Ty. 1, Cl. 1[c]	V747-75
NAS1596	MIL-R-25897, Ty. 1, Cl. 2[c]	V-709-90
NAS1612	NAS1613	E515-80

1	2	3	4	5	6	7	8	9		10	11	12	13	
					Metric O-ring size (mm)					O-ring size—actual[e] per ARP 568 (in.)				
						Tolerance[f]					Tolerance[f]			
	Sequence A and AS568	Sequence B	Sequence C	Tube OD		Class I	Class II	w			Class I	Class II	w	
3-XX[d] size no.	dash no.	dash no.	dash no.	(ref.)	ID	±	±		±	ID	±	±		±
3-901	−901	−01		3/32	4.70	.13	.15	1.42	.08	.185	.005	.006	.056	.003
3-902	−902	−02	−2	1/8	6.07	.13	.18	1.63	.08	.239	.005	.007	.064	.003
3-903	−903	−03	−3	3/16	7.65	.13	.18	1.63	.08	.301	.005	.007	.064	.003

3-904	−904	−04	−4	1/4	8.92	.13	.18	1.83	.08	.351	.005	.007	.072	.003
3-905	−905	−05	−5	5/16	10.52	.13	.18	1.83	.08	.414	.005	.007	.072	.003
3-906	−906	−06	−6	3/8	11.89	.13	.18	1.98	.08	.468	.005	.007	.078	.003
3-907	−907	−07		7/16	13.46	.13	.18	2.08	.08	.530	.005	.007	.082	.003
3-908	−908	−08	−8	1/2	16.36	.13	.23	2.21	.08	.644	.005	.009	.087	.003
3-909	−909	−09		9/16	17.93	.13	.23	2.46	.08	.706	.005	.009	.097	.003
3-910	−910	−10	−10	5/8	19.18	.13	.23	2.46	.08	.755	.005	.009	.097	.003
3-911	−911	−11		11/16	21.92	.13	.23	2.95	.10	.863	.005	.009	.116	.004
3-912	−912	−12	−12	3/4	23.47	.15	.23	2.95	.10	.924	.006	.009	.116	.004
3-913	−913	−13		13/16	25.04	.15	.25	2.95	.10	.986	.006	.010	.116	.004
3-914	−914	−14	−14[g]	7/8	26.59	.15	.25	2.95	.10	1.047	.006	.010	.116	.004
3-916	−916	−16	−16	1	29.74	.15	.25	2.95	.10	1.171	.006	.010	.116	.004
3-918	−918	−18	−18	1-1/8	34.42	.15	.30	2.95	.10	1.355	.006	.012	.116	.004
3-920	−920	−20	−20	1-1/4	37.47	.25	.36	3.00	.10	1.475	.010	.014	.118	.004
3-924	−924	−24	−24	1-1/2	43.69	.25	.36	3.00	.10	1.720	0.10	.014	.118	.004
3-928	−928	−28	−28	1-3/4	53.09	.25	.46	3.00	.10	2.090	0.10	.018	.118	.004
3-932	−932	−32	−32	2	59.36	.25	.46	3.00	.10	2.337	0.10	.018	.118	.004

*AND 10049 and AND 10050 were canceled Dec. 14, 1966.

aSpecification MIL-R-25988 requires special documentation.

bNonstandard compound. Made to order only.

cSpecification inactive for new design.

dThe rubber compound must be added when ordering by the 3- size number (i.e., 3-910 N552-90).

eMaterial with unusual shrinkage during molding will give slightly different dimensions.

fClass II tolerances apply to columns 13 and 14 (M83248/1 and M83248/2) and to M25988/1, M25988/3, NAS1593, NAS1594, NAS1595, and NAS1596. However, AS568 A, revision A, established a single set of ID tolerances. This was agreed on by the Air Standardization Committee (membership by the United States, Australia, Canada, New Zealand, and the United Kingdom).

gMS28778 only.

Source: O-Ring Handbook OR5700, Parker Seal Co., Lexington, Ky., January 1977.

Table 5. O-Rings for Electrical Connectors

MS28900[a]	Parker Size No.	ID	±	W	±
− 8	5-133	0.332	0.005	0.031	0.003
−10	5-134	0.410	0.005	0.031	0.003
−12	5-135	0.526	0.005	0.031	0.003
−14	5-136	0.643	0.005	0.031	0.003
−16	5-137	0.775	0.006	0.031	0.003
−18	5-138	0.898	0.006	0.031	0.003
−20	5-139	0.987	0.006	0.031	0.003
−22	5-140	1.112	0.006	0.031	0.003
−24	5-141	1.226	0.006	0.031	0.003
−28	5-142	1.450	0.010	0.047	0.003
−32	5-143	1.670	0.010	0.047	0.003
−36	5-144	1.891	0.010	0.047	0.003

[a]Rubber material for MS28900 O-rings—Parker Compound C557-70 (Specification AMS3209).
Source: O-Ring Handbook OR5700, Parker Seal Co., Lexington, Ky., January 1977.

standard on which AS568 was developed. M25988, M83248, MS9020, MS9355, and MS29512 cover 31 O-rings of slightly larger diameter cross sections to be used for sealing straight-thread tube fittings. These O-rings are usually referred to as "3 dash 9" O-rings, as shown in Table 4, and are equivalent to dash numbers 901 through 932 of AS568. They are used in gland designs specified by military standards MS16142, MS33649, and MS33656 for tube fittings. Table 5 presents the usual seals used in gland designs for electrical connectors under MS28900. The specific dimensions for the glands required by these O-ring sizes are presented under the Specific Applications of Static and Dynamic Seals.

Most O-ring sizes have been specified such that the dimensions of the gland in which they will be installed meet standard machine tooling. Therefore, a female gland seal will have a standard bore diameter, requiring common tooling so as not to increase machining costs.

III. COMPOUNDS AND MATERIALS

Elastomeric O-ring seals are generally composed of at least two monomers linked end to end to form long-chain molecules. The

Table 6. Comparison of Properties of Commonly Used Elastomers[a]

Property	Nitrile or Buna-N (N)	SBR or Buna-S (G)	Butadiene (D)	Butyl (B)	Neoprene (C)	Chlorosulfonated Polyethylene (H)	Ethylene Propylene (E)	Fluorocarbon (V)	Fluorosilicone (L)	Isoprene (I)	Natural Rubber (R)	Polyacrylate (A)	Polysulfide (T)	Polyurethane (P)	Silicone (S)	Epichlorohydrin (Y)
Ozone resistance	P	P	P	GE	GE	E	E	E	E	P	P	E	E	E	E	E
Weather resistance	F	F	F	GE	E	E	E	E	E	F	F	E	E	E	E	E
Heat resistance	G	FG	F	GE	G	G	E	E	E	F	F	E	P	F	E	E
Chemical resistance	FG	FG	FG	E	FG	E	E	E	E	FG	FG	E	P	F	GE	FG
Oil resistance	E	P	P	P	FG	P	P	E	G	P	P	E	E	G	PG	G
Impermeability	G	F	F	E	G	F	G	E	P	F	F	E	E	G	E	E
Cold resistance	G	G	G	G	FG	G	GE	P	F	G	G	P	G	G	E	GE
Tear resistance	FG	FG	GE	G	FG	FG	GE	FP	P	GE	GE	FG	P	GE	P	GE
Abrasion resistance	G	G	E	G	G	G	GE	G	P	E	E	G	P	E	P	G
Set resistance	GE	G	G	FG	F	F	GE	GE	P	G	G	P	P	E	GE	PF
Dynamic properties	GE	G	F	F	F	F	GE	GE	P	G	G	F	P	F	GE	G
Acid resistance	F	F	FG	G	FG	G	G	GE	E	F	E	F	F	E	P	G
Tensile strength	GE	F	FG	G	FG	F	G	E	FG	E	E	F	F	E	FG	FG
Electrical properties	F	G	E	F	F	G	G	GE	E	G	E	F	F	FG	E	G
Water/steam resistance	FG	FG	FG	G	F	F	E	FG	F	FG	FG	P	F	P	F	F
Flame resistance	P	P	P	P	G	G	P	E	G	P	P	P	P	P	F	FG

Key: P, poor; F, fair; G, good; E, excellent

[a]See Charts 1A–1D and Table 11 (pp. 55–63) for additional information.

Source: O-Ring Handbook OR5700, Parker Seal Co., Lexington, Ky., January 1977.

molecules are in turn linked together by the vulcanization process.
Vulcanization determines the number of molecules to be linked together: this determines the strength and elasticity of the seal material. The chemical structure of the monomer in the molecular
chain determines the material's resistance to deteriorating influ-
ences such as heat, cold, oils, solvents, and other chemicals.
Many catalysts and compounds may be added to the basic monomers
to affect either the strength or chemical characteristics of the
final material. Such additives as curing agents, accelerators,
fillers, reinforcing carbon blacks, process aids, antioxidants, and
plasticizers are introduced into the mixing and vulcanization process at specific times and temperatures to influence the finished
product.

Table 6 presents a comparison of properties of the most used
elastomeric materials. This table gives a good indication of the
various limitations of each material, but applications involving
combinations of chemical, oil, and heat resistance require further
investigation. For example, ethylene propylene has an excellent
rating under chemical resistance, while nitrile or Buna-N has a
fiar to good rating. But if the particular application calls for
sealing a chemical fluid that also contains a mineral oil ingredient,
the ethylene propylene seal will fail (dissolve), while the nitrile
seal would be acceptable. Of course, in this case, a fluorocarbon
seal would be the first choice, if available. Many seal sizes and
configurations are not readily available in all estomeric materials.

The most common elastomeric materials are nitrile or Buna-N,
butyl, neoprene, and ethylene propylene. These, and each of
the other materials shown in Table 6, are discussed below.

1. Nitrile or Buna-N Rubber: More nitrile seals are used than
all the other elastomers combined, since nitrile is the most versatile material. Nitriles are a copolymer of acrylonitrile and butadiene [1, p. 60]. As the acrylonitrile content of nitriles increases,
the oil and fuel resistance increases while the low-temperature
flexibility decreases. Nitrile-base elastomers are usually specified
by military MS and AN O-rings when used in oil and fuel applications, but because nitrile compounds vary widely within such a
large overall temperature range, particular attention should be
paid to specifying physical properties. Materials can be formulated
to perform satisfactorily over the temperature range -65 to $+300°F$,
so it is necessary to make sure that the particular nitrile chosen
meets the temperature requirements of the application (see Fig. 2).

The nitrile materials are recommended for general-purpose
sealing of alkaline and salt solutions, petroleum oils and fluids,
vegetable and diester oils, silicone greases and oils, ethylene

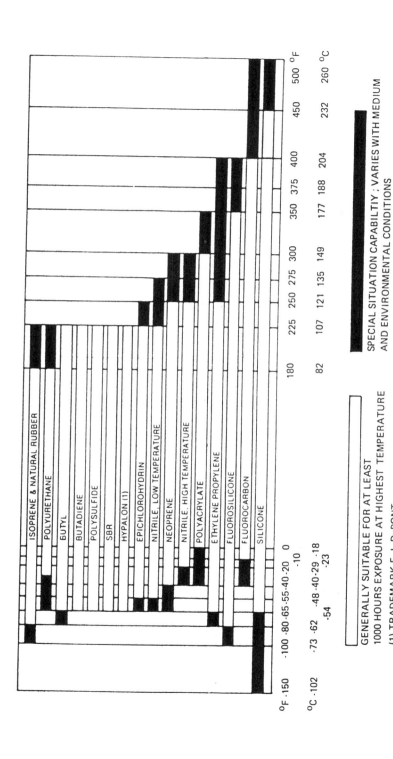

Figure 2. Temperature capabilities of principal elastomers employed in seals (*From O-Ring Handbook, OR5700,* Parker Seal Co., Lexington, Ky., January 1977)

glycol-base fluids, alcohols, gasolines, and water. They are not suited for use with strong oxidizing agents; chlorinated solvents such as carbon tetrachloride or trichlorethylene, nitrated hydrocarbons such as nitrobenzene or aniline; phosphate esters such as Skydrol, Fyrquel, or Pydraul; acetates; keytones such as methyl ethyl ketone (MEK) and acetone; and aromatic hydrocarbons. Ozone will usually attack nitrile materials, but resistance can be greatly improved by the addition of antiozonant compounds.

2. *SBR or Buna-S*: Although styrene-butadiene rubber (SBR) and natural rubber account for approximately 90% of the world's total rubber consumption, very little of these two materials is used in seals. SBR is a copolymer of styrene and butadiene and has been referred to as government rubber styrene (GRS) since it was developed during World War II. The butadiene/styrene ratio determines the low-temperature characteristics of the material.

SBR seals have been used in automotive brake systems, in systems employing alcohols and water, and to a limited extent in systems exposed to nitroglycerides. The material functions within a temperature range common to other natural and early synthetic rubber products such as butadiene and polysulfide: −65 to +255°F. SBR is not recommended for petroleum oils, or in systems exposed to sunlight.

3. *Butadiene Rubber*: Polybutadiene is an elastomer with physical properties slightly less than those of natural rubber. Its low-temeprature characterists have been improved over natural rubber by special compound additives. Butadiene is used primarily by the tire industry.

4. *Butyl Rubber*: Butyl rubber is a petroleum material made by compounding isobutylene and isoprene. Brominated and chlorinated butyl rubber are also available and are prepared by select replacement of hydrogen with bromine or chlorine [2, p. A3-3]. These materials have very good resistance to ozone, vegetable oils, some mineral acids, ketones, phosphate ester hydraulic fluids such as Skydrol, Fryquel, and Pydraul, and silicone fluids and greases. Butyl is not recommended for petroleum oils or diester-base lubricants. The elastomeric industry is currently recommending the use of ethylene propylene rubber instead of butyl for phosphate ester application.

Butyl rubber has excellent resistance to gas permeation and is extensively used in vacuum chambers and gas containers. Its ability to absorb energy in vibration-damping applications has led to its use in isolation mounts and dynamic stop systems.

5. *Neoprene Rubber of Chloroprene*: Neoprene rubber is one of the earliest synthetic materials used by the seal manufacturers.

It is a homopolymer of chloroprene possessing resistance to weather elements of ozone, oxygen, and sunlight. Unlike most elastomers that either resist weather or petroleum products, neoprene has limited resistance to both. Because of this and its rather broad temperature range, it is the usual choice in compromising applications.

Neoprene materials are used to seal dilute acids, bases and salts, straight-chain hydrocarbons, high-aniline-point petroleum oils, vegetable oils, alcohols, and silicate ester lubricants. Neoprene is used extensively with refrigerants of Freon and ammonia, but is unsatisfactory for service in aromatic hydrocarbons, chlorinated solvents, esters, and ketones.

6. *Chlorosulfonated Polyethylene (CSM)*: As the name implies, this polymer is produced by compounding polyethylene with chlorine and sulfur dioxide. The chemical reaction of chlorine and sulfur dioxide transforms the thermoplastic polyethylene into a synthetic rubber. The addition of curing agents and accelerators in the vulcanization process produces elastomers that have excellent resistance to ozone, weather, and oxidizing chemicals. Chlorosulfonated polyethylene (trade name Hypalon) has only fair set-resistance qualities and is therefore not recommended for dynamic seal applications.

7. *Ethylene Propylene Rubber (EPM, EPDM, EPR)*: This elastomer is a copolymer of ethylene and propylene and is sometimes compounded with a third monomer (EPT). Good to excellent compression set resistance is obtained by the addition of peroxide cures during vulcanization. Ethylene propylene materials have excellent resistance to phosphate esters such as Skydrol, Fyrquel, Pydraul, water and steam, acids, alkali, salt solutions, ketones, alcohols, glycols, and silicone oils and greases. EPR has very poor resistance to petroleum oils and diester-base lubricants. Ethylene propylene is a close contender to Buna-N and butyl in the important sealing properties, except that it does not have the petroleum oil and fuel resistance of Buna-N, nor the low-gas-permeability quality of butyl [1, p. 61].

8. *Fluorocarbon Rubber (FKM)*: Fluorocarbon elastomers have been compounded to meet a wide range of chemical and physical requirements. Under the trade names Viton, Fluorel, and Kel-F, fluorocarbon seals have been employed where other materials cannot survive severe chemical conditions. The working temperature range of FKM is between -20 and $+400°F$ (-29 and $+204°C$) and limited temperature spikes of 600°F have been incurred. New compoundings have greatly improved the compression set of fluorocarbon O-ring seals.

Fluorocarbon elastomers have good resistance to the swelling
and deteriorating effects of aromatic solvents, aliphatic hydrocarbons, halogenated hydrocarbons such as carbon tetrachloride,
trichloroethylene, diester oils, silicate ester oils, petroleum oils,
and many mineral acids. They are also highly recommended in
applications involving ozone combined with heat, as in electric
motors and electrical equipment. Fluorocarbons are not recommended for use with highly polar fluids such as hydrazine, ketones, phosphate esters like Skydrol, and anhydrous ammonia,
and are also not recommended with low-molecular-weight esters
and hot hydrofluoric or chlorosulfonic acids.

9. *Fluorosilicone (FSi, SLS)*: Fluorosilicone polymers are
made by replacing the methyl side groups on the silicone polymer
chain with fluorinated side groups. This produces a material
risistant to hydrocarbon fuels and oils at both high and low temperatures. The primary uses of fluorosilicones are in fuel systems
at temperatures up to 350°F (177°C), and in applications where
the dry-heat resistance of silicone is required, but the seal may
be exposed to petroleum oils and/or hydrocarbon fuels [2, p.
A3-5]. Because of fluorosilicone's poor dynamic properties,
fluorosilicone seals are used only for static seal applications.
There is a current effort in the seal industry to color-code fluorosilicone seals blue to distinguish them from the normally colored
red silicone seals.

10. *Isoprene Synthetic Rubber (IR, SN)*: Polyisoprene materials are equivalent in chemical and physical properties to
natural rubber, except for some limitations when used in dynamic
applications. Isoprene seals have excellent tensile strength and
abrasion resistance, but because isoprene has only fair dynamic
properties, it is not recommended for rotary or reciprocating seal
applications.

11. *Natural Rubber (NR)*: The majority of natural rubber
used today is produced by the *Hevea brasiliensis* tree found in the
Far East and Brazil. Nautral rubber has excellent physical resistance to abrasion, tension, dynamic distortion, and cold flow.
Natural rubber seals are currently being replaced by synthetics
that can be compounded to better resist sunlight, oxygen, ozone,
solvents, and oils. Natural rubber seals are still used for sealing
many automobile hydraulic brake systems.

12. *Polyacrylate (ACM)*: There are several polymer products
of acrylic acid esters, common trade names being Hycar, Krynac,
Thiacril, and Cyanacryl. These materials maintain excellent resistance to the weathering elements of sunlight, oxygen, and ozone,
even under severe flexual distortion. These materials also possess

high resistance to petroleum fuels and oils, but very poor chemical resistance. Polyacrylate seals are used primarily in automobile engines and transmissions where hot oils are incurred, temperature capability ranging from 0 to 350°F (−18 to +177°C).

13. *Polysulfide Rubber (Thiokol)*: This material is not normally recommended for O-ring seals because it has very poor physical characteristics. It is only recommended when the special case of low-temperature flexibility and crack resistance is required in the presence of solvents or weathering elements of ozone, oxygen, and sunlight. These solvents may include ketones, ethers, and petroleums.

14. *Polyurethane (AU,EU)*: Polyurethane elastomers are compounds of polyethers and diisocyanates. These materials have excellent physical properties of abrasion resistance and tensile strength, which make them outstanding for dynamic applications. They have excellent resistance to weather, ozone, and oxygen, good resistance to hydrocarbon fuels, petroleum oils, and aliphatic solvents, and fair resistance to aromatic hydrocarbons. Acids, ketones, and chlorinated hydrocarbons attack and deteriorate polyurethane. Because polyurethane is available in castable liquids, injection-moldable pellets, and millable gums, it is a very useful material for unique and specialized sealing problems.

15. *Silicone Rubber (Si)*: Silicone elastomers are compounded from dimethyl silicone polymers, and thus will deteriorate if used with silicone oils or greases. Various additives have extended the functional temperature range of silicone rubber beyond any other elastomer. Flexibility below −175°F (−114°C) and service above 700°F (371°C) for short periods of time have been demonstrated [2, p. A3-5]. High production costs have normally limited the use of silicone seals to applications requiring extreme temperature resistance. Production molding of silicone seals involves high-temperature secondary cure which results in greater than normal shrinkage. The finished O-ring seal is usually undersized when produced in standard molds. The designer should be aware of this size difference when designing glands for silicone O-rings.

Silicone elastomers have poor resistance to ketone solvents such as MEK and acetone, and poor resistance to most petroleum fluids. They have very poor physical properties that make them unattractive for dynamic applications. Silicone seals are recommended for extreme temperature use with ozone, oxygen, high-aniline point oils, and chlorinated diphenyls.

16. *Epichlorogydrin Rubber (CO,ECO)*: These chloroaliphatic polymers are oil-resistant materials compounded in two distinct

forms, homopolymers and copolymers. The functional temperature range of homopolymers is $-40°F$ to $+275°F$ ($-40°C$ to $135°C$),
while that of copolymers is $-65°F$ to $+275°F$ ($-54°C$ to $135°C$).
Both have excellent resistance to ozone, weather, and hydrocarbon oils and fuels. The corrosive and poor set-resistance properties of epichlorogydrin have limited its use in some seal applications.

Figure 2 presents the temperature capabilities of the principal
elastomeric materials discussed above. The temperature limits
given include a realistic safety margin dependent on the most
common compounding of the specific material. The maximum-tem-
perature recommendation for a material is based on reliable functional service for 1000 hr. Time at less than maximum temperature
will extend life, while higher temperature will reduce life. The
low-temperature service limit is based on the TR-10 value per
ASTM D1414 minus a nominal 15°F. This method ensures reliable
low-temperature service, provided that sufficient squeeze is designed into the gland geometry. It has been determined that
elastomeric materials which can withstand extreme compressive
forces are good for low-temperature seal applications. The extreme compressive forces required are those encountered when
the material can be compressed to less than 20% of its original
thickness without damage [1, p. 5.].

The actual seal environment may extend beyond the temperature limits specified in Fig. 2 for short periods of time. For example, a fluorocarbon seal which is limited to 400°F for prolonged
service can tolerate 2000°F for 20 to 40 sec without damage. It is
also important to note that no permanent damage is done to any
elastomeric material by cryogenic temperatures. A frozen O-ring
will regain its service characteristics when returned to normal
operating temperatures.

IV. ELASTOMERIC SPECIFICATIONS

Table 7 presents the commonly referenced military and aerospace
specifications for the most popular elastomeric materials. The
common name of the material is given together with a brief des-
cripetion of its featured characteristic and/or use. This table
should be useful in categorizing and cross-referencing the prom-
inent military and aerospace specifications. The designer should
be careful not to select one of these military or aerospace specifications without being sure that it is broad enough to cover the intended use of the material capabilities, and on the other hand,

Table 7. Elastomeric Specifications: Military Aerospace Material Specification: National Aerospace Standard

Common name	Specification	Description
Nitrile (Buna-N)	MIL-G-1149: Type 1 and 2 Class 5	Gasket materials, synthetic rubber
	MIL-R-3533	Rubber, synthetic; sheet, strip, and molded
	MIL-P-5315	O-ring packing, hydrocarbon fuel resistant
	MIL-P-5510	Gasket, straight-thread tube fitting boss
	MIL-P-5516: Class B	Gasket and packing, hydraulic, aircraft
	MIL-R-6855: Class 1, Grade 60; Class 2, Types A and B	Synthetic sheets, strips, molded, extruded shapes
	MIL-R-7362: Types 1 and 2	Synthetic sheet, molded and extruded shapes
	MIL-G-21569: Class 1	Gasket cylinder linear seal, synthetic rubber
	MIL-G-21610: Type 1	Gasket, heat exchanger, synthetic rubber
	MIL-G-23983	Gasket, packing material, oil resistant
	MIL-P-25732	Preformed packing, petroleum fluid resistant
	MIL-P-83461	Preformed packing, petroleum fluid resistant
	AMS3201, AMS3202	Dry heat resistance
	AMS3205	Low temperature resistance
	AMS3212, AMS3215	Aromatic fuel resistant
	AMS3220	General purpose; fuel, oil, glycol
	AMS3226, AMS3227, AMS3228	Hot oil and coolant resistant, low swell
	AMS3229	Hot oil resistant, low swell
	AMS7260, AMS7271	Fuel and low temperature resistant
	AMS7270	Fuel resistant
	AMS7272	Synthetic lubricant resistant
	AMS7274	Oil resistant

27

Table 7. (continued)

Common name	Specification	Description
SBR or Buna-S	MIL-G-1149: Type 2, Class 2	Gasket material, synthetic rubber
Butyl	AMS3238 AMS7277	Phosphate ester resistant, butyl type Phosphate ester, hydraulic fluid resistant
Neoprene	MIL-G-1149: Types 1 and 2, Class 1 AMS3208, AMS3209, AMS3240, AMS3241, AMS3242 AMS3222	Gasket material, synthetic rubber Weather resistant, chloroprene type Hot oil resistant, high swell
Ethylene	MIL-G-22050 MIL-R-83285 NAS1613	Synthetic rubber, nonflammable hydraulic fluid General purpose, ozone resistant Packing, O-ring, phosphate ester resistant
Fluorocarbon	MIL-G-23652: Types 1 and 2 MIL-R-25897: Type 1, Classes 1 and 2 MIL-R-83248: Type 1, Classes 1 and 2	Gasket and packing material, petroleum and phosphate ester fluid resistant Rubber, high temperature fluid resistant Rubber, fluorocarbon, high temperature fluid and compression set resistant

MIL-R-83485: Type 1	Rubber, fluorocarbon, low temperature performance
AMS7276	High temperature fluid resistant, very low compression set fluorocarbon
AMS7278, AMS7279	High temperature fluid resistant fluorocarbon
AMS7280	Low compression set fluorocarbon
Fluorosilicone	
MIL-R-25988: Type 1, Class1	Rubber, fluorosilicone, oil, and fuel resistant
AMS3326	Silicone, fuel, and oil resistant
Silicone	
MIL-R-5847 (superseded by ZZ-R-7658)	Rubber silicone
MIL-G-21569: Class 2	Gasket cylinder linear seal, synthetic rubber
AMS3301, AMS3302, AMS3303, AMS3304, AMS3305, AMS3307	Silicone, general purpose
AMS3335, AMS3337	Silicone, extreme low temperature resistant
AMS3345, AMS3349	Silicone rubber
AMS3357	Silicone rubber, lubricating oil, compression set resistant
AMS7267	Silicone, heat resistant, low compression set

that it is not so broad that it allows variations in material compound that may not be acceptable.

V. MANUFACTURERS AND DISTRIBUTORS

Table 8 lists 49 manufacturers and/or distributors of elastomeric ring seals according to location within the United States. This list is included to aid the designer in obtained additional information directly from a local manufacturer or distributor. This is not an all-inclusive list.

Table 8. Elastomeric Ring Seals: Manufacturers and Distributors[a]

Alabama
  Sepco Corp., Birmingham
  Tenn-Val Inc., Decatur

California
  Satori Seal Corp., Alhambra
  Gasket Specialties Inc., Brisbane
 *Kyowa Metriseal Co., Brisbane
  General Connectors Corp., Burbank
  Standard Polymer, Inc., Camarillo
  Arvan, Inc., El Monte
  Bozung, J. A. Co., El Monte
  Allmetal Screw Products Company, Inc., El Segundo
  Tetrafluor, Inc., El Segundo
  Bobber Products, Inc., Fullerton
  Aero-Stat Co., Gardena
  Porter Seal Mfg., Glendale
 *Advantec, Inglewood
  Service Rubber & Gasket Co., La Puente
  Holz Rubber Co., Lodi
  Fluorocarbon Co., Los Alamitos
  Airsco, Los Angeles
 *Burly Seal Products Co., Los Angeles
  Calnevar Sea Co., GSC Corp., Los Angeles
  Local Co., Los Angeles
  Majestic Fasteners Co., Los Angeles
  Material Fabricators, Inc., Los Angeles
  O-Rings Inc., Los Angeles
  Plastic and Rubber Engineering, Inc., Los Angeles
  Alatec Products, North Hollywood

Table 8. (continued)

*Parco, Ontario
Dodge-Wasmund Mfg., Inc., Pico Rivera
Diesel Energy Products, Inc., Pleasanton
Vibration Isolation Products, Inc., San Fernando
American Asbestos Co., San Francisco
Western Rubber & Supply, San Francisco
Burke Industries, San Jose
Boyd Industries, San Leandro
Bal-Seal Engineering Co., Santa Ana
International Seal Company, Inc., Santa Ana
Allmetal Screw Products Company, Inc., Santa Clara
Qualified Air Components, Santa Fe Springs
Sealtec Inc., Santa Fe Springs
Allied Airparts & Supply Company, Inc., South El Monte
*Houston Rubber Company, Inc., Sylmar
*Da/Pro Rubber, Inc., Van Nuys

Connecticut
Agi Rubber Company, Bridgeport
Enflow Corp., Bristol
*Connecticut Rubber Molding Corporation, Danielson
Magnatec, East Ganby
Parts Inc., East Hartford
American Seal & Engineering Company, Inc., Hamden
Agc Inc., Meriden
Auburn Mfg. Company, Middletown
Advance Products Company, Inc., North Haven
Orcomatic Inc., Norwich

Florida
Hoose, Robert E., Inc., Miami
Veri-Tech Inc., Pompano Beach
Century Fasteners Corp., Tampa
Allmetal Screw Products Company, Inc., Winter Park

Georgia
AAA Seals & Packing Company, Inc., Atlanta
Allmetal Screws Products Company, Inc., Atlanta
Dixie Bearings, Inc., Atlanta
Dixie Packing Company, Atlanta
General Rubber & Plastics Corp., Atlanta

Table 8. (continued)

Illinois
  TFE-O-SIL Corp., Addison
  Nok-Inc., Bensenville
  Woods Mfg. Company, Inc., Bensenville
  Aero Rubber Company, Inc., Bridgeview
  Bostik/Stalok, Broadview
  Acadia, Chicago
  Accurate Products Inc., Chicago
  Atlantic India Rubber Company, Chicago
  Chicago Gasket Company, Chicago
  Flow Products Inc., Chicago
  Simrit Corporation, Des Plaines
  Hologen Insulator & Seal Corp., Elk Grove
  Lutz Sales Company, Inc., Elk Grove Village
  Quality Rubber Mfg. Company, Inc., Elk Grove Village
  Allmetal Screw Products Company, Inc., Elmhurst
  Anchor Bolt & Screw, Melrose Park
  Crane Packing Company, Morton Grove
 *Gil-Bar Rubber Products Company, Oak Park
  SBI Incorporated, Tool Division, Peoria
  Amtrex Corporation, Rockford
 *Excelsior Incorporated, Rockford

Indiana
 *Continental Seal Corporation, Fort Wayne
  Hallite Seals Inc., Fort Wayne
  Press-Seal Gasket Corp., Fort Wayne
  Seals Division, W. S. Shamban & Company, Fort Wayne
 *Goshen Rubber Company, Goshen
  T & M Rubber Incorporated, Goshen
 *Tri-Seals Incorporated, Goshen
  Triangle Rubber Company, Goshen
 *Ligonier Rubber Company Inc., Ligonier
  New Castle Engineering Inc., New Castle

Iowa
  Octa-Ring Seal Company, Beaman

Kentucky
 *Gayle, Geo. W., & Son, Frankfort
  Parker Seals O-Ring Division, Lexington
  Moore I. B. Corporation, Lexington

Table 8. (continued)

Maryland
   Paramount Packing & Rubber Inc., Baltimore
   Phelps Packing & Rubber Co., Inc., Baltimore
   *Pressure Science Inc., Beltsville
   Allmetal Screw Products Company, Inc., Ellicott City
   Maryland Metrics, Owings Mills

Massachusetts
   Metrics For Industry, Avon
   *Orion Industries, Inc., Bolton
   *Moore, Irvin B. Corporation, Cambridge
   *Paul-Martin Rubber Company, Holyoke
   Allmetal Screw Products Company, Inc., Waltham
   Shelly, John G. Company, Inc., Wellesley Hills
   Rush Industries Inc., Winchester

Michigan
   *Hoover Universal, Inc., Foam Division, Ann Arbor
   *Hoover Universal Inc., Plastic Components Division, Ann Arbor
   *Seal Comp Inc., Brighton
   Federal-Mogul Corporation, Detroit
   Federal-Mogul Corporation, Industrial Sales, Detroit
   *Industrial Plastics & Mfg. Company, Detroit
   *Mold-Ex Rubber Company, Inc., Farmington
   *Uracast Products Inc., Fenton
   *Zefflamb Industries Ioc., Fenton
   Allmetal Screw Products Company, Inc., Ferndale
   *Jackson Flexible Products Inc., Jackson
   Shurclose Seal Company, Lake Orion
   Mather Company, The, Fluorotec Division, Milan
   March, C. W. Company, Muskegon
   *Galap Rubber Incorporated, Plymouth
   L & L Products Incorporated, Romeo
   *Way Wipers Incorporated, Royal Oak
   Accro-Seal Incorporated, Vicksburg
   *Uniflex Incorporated, Wixom

Minnesota
   Circle Rubber Company, Eden Prairie
   Minnesota Rubber Company, Minneapolis
   *Molding Technical Systems Incorporated, Minneapolis
   Precision Associates Inc., Minneapolis
   *Robinson Rubber Products, Minneapolis

Table 8. (continued)

Interchange Incorporated, St. Louis Park
Rubber Industries Incorporated, Shakopee

Missouri
Hanna Rubber Company, Kansas City
Gasket & Seal Fabricators, St. Louis

New Hampshire
Frederickseal Incorporated, Bedford
Disogrin Industries Corporation, Manchester

New Jersey
Eastern Molding Company, Belleville
Jet Engine Supply Company, Bellmawr
Abesto Corporation, Bloomfield
*Minor Rubber Company, Inc., Bloomfield
*Helicoflex Company, Boonton
*Star-glow Industries Incorporated, East Rutherford
Industrial Rubber Company, Elizabeth
Stevens Associates Incorporated, Emerson
APM-Hexseal, Englewood
Conover, C. E. & Company, Inc., Fairfield
*Multi-Flex Seals, Incorporated, Hackensack
Hawthorne Rubber Mfg. Corporation, Hawthorne
AME Corporation, Little Falls
Princeton Rubber Company, Inc., Monmouth Junction
Federal Carbine Company, Moonachie
Janos Industrial Insulation Corporation, Moonachie
Goodyear Rubber Products Corporation, Newark
Astro Molding Incorporated, Old Bridge
Unette Corporation, Parsippany
Alatec Products, Pine Brook
Seals Eastern Incorporated, Red Bank
Sterling Plastics & Rubber Products Incorporated, South Amboy
Chemplast, Incorporated, Wayne
Miller, Franklin, West Orange
Sea-Ro Packing Company, Incorporated, Wood-Ridge

New York
Hercules Products, Alden
Balfor Industries Incorporated, Bronx
Emerson Plastronics Incorporated, Bronx
Tri-Component Products Corporation, Bronx
Patterson Machine Company, Brooklyn

Table 8. (continued)

Great Lakes Plastic, Buffalo
Sevatronics Incorporated, Buffalo
Berg, Winfred M. Incorporated, East Rockaway
Century Fasteners Corporation, Elmhurst
Enbee Mfg. Company, Incorporated, Elmont
Girard Rubber Corporation, Elmsford
*Northeast International Corporation, Elmsford
Technical Specialties Company, Incorporated, Elmsford
S. A. S. Gasket and Supply Company, Incorporated,
    Farmingdale
Allmetal Screw Products Company, Inc., Garden City
Millimeter Industrial Supply Corporation, Hauppauge
Metric & Multistandard Components Corporation, Hawthorne
Fluorglas Division, Oak Materials Group, Incorporated, Housick
    Falls
Nationwide Metric Supply Corporation, Huntington Station
Apple Rubber Products Incorporated, Lancaster
*Sealing Devices, Lancaster
U. S. Rubber Supply Company, Long Island City
Product Components Corporation, Mount Vernon
*Columbia Nut & Bolt Corporation, New York
Presray Sub of Pawling Rubber Corporation, Pawling
Gaddis Engineering Company, Port Washington
Allmetal Screw Machine Products Company, Inc., Rochester
Smith Rubber Company, Inc., Rochester
Action Associates, Roslyn Heights
*Minisink Rubber Company, Inc., Unionville
Northeast International Corporation, White Plains

North Carolina
Allmetal Screw Products Company, Inc., Charlotte
Century Fasteners Corporation, Charlotte
Blue Ridge Division, Lavren Mfg. Company, Fletcher
CGR Products, Greensboro

Ohio
Ohio Gasket & Shim Company, Inc., Akron
Qualiform Incorporated, Barkerton
Allmetal Screw Products Company, Incorporated, Cincinnati
Sur-Seal Gasket & Pasking Incorporated, Cincinnati
Alan International Incorporated, Cleveland
Allmetal Screw Products Company, Incorporated, Cleveland
Bearings Incorporated, Cleveland

Table 8. (continued)

Bruening Bearings Incorporated, Cleveland
Durox Equipment Company, Cleveland
Flex Incorporated, Cleveland
Merkel-Forsheda Corporation, Cleveland
Grotenrath Rubber Products, Cleveland
Parker-Hannifin Corporation, Cleveland
Universal Grinding Corporation, Cleveland
C & M Rubber Company, Dayton
Continental Tool & Rubber Products Incorporated, Dayton
Rubbercraft Incorporated, Dayton
Rubber-Tech Incorporated, Dayton
*Fluorocarbon-Sparta, Dover
High Quality Plastics, Findlay
*Hoover Universal Incorporated, Mansfield
Merg, Mantua
EGC Enterprises Inc., Mentor
Forest City Products, Incorporated, Mentor
Polymer/Raymond Industries Incorporated, Middlefield
AMG Industries Incorporated, Mount Vernon
*Profile Rubber Corporation, Sharon Center
Standard Carbon Company, Steubenville
Urethane Products Industries Incorporated, Stow
*Industrial Electronics Rubber Company, Twinsburg
Industrial Packing & Seals Incorporated, Uniontown
Forest City Foam Products Incorporated, Wellington
Flite Hardware, Westlake
Vernay Laboratories Incorporated, Yellow Springs

Oregon
Huntington Rubber Company, Portland

Pennsylvania
Pars Mfg. Company, Ambler
*Dooley, James E. Company, Broomall
Castle Rubber Company, Butler
*Clifton Plastic Incorporated, Clifton Heights
Allegheny Plastics Incorporated, Coraopolis
*Weaver Industries Incorporated, Denver
Drummond Rubber Products, Doylestown
Reliable Rubber Products Company, Eddington
Beemer Engineering Company, Fort Washington
Industrial Gasket & Shim Company, Inc., Meadowlands
Lehigh Rubber Works Incorporated, Morrisville

Table 8. (continued)

Greene, Tweed & Company, North Wales
Anchor Packing Company, Philadelphia
*Apex Molded Products Company
Allen Stevens Electrical Fittings Company, Scranton
Atlas Technology Incorporated, Scranton
Louis H. Hein Company, West Conshohocken
Allmetal Screw Products Company, Willow Grove
*DBR Industries Incorporated, Yardley
B & D Supply Incorporated, Yeadon
Allegheny-York Company, York

Rhode Island
Dixon Industries Corporation, Bristol
EG & G Sealol Incorporated, Providence

South Carolina
UAP Components Incorporated, Columbia
J-B-L Division of Parker Seals, Spartanburg
Reeves Brothers Incorporated, Spartanburg
Hartwell Plastics Incorporated, Starr

Tennessee
Mountain Empire Rubber & Specialty Company, Inc., Johnson
City
Century Fasteners Corporation, Knoxville
Precision Rubber Products Corporation, Lebanon

Texas
Allmetal Screw Products Company, Incorporated, Dallas
Oriental Gasket & Packing Company, Dallas
Gorman Company Incorporated, Duncanville
Butler Professional Services Group Incorporated, Highlands
All-Seal-Texas Incorporated, Houston
K & W Incorporated, Houston
Maloney, F. H. Company, Houston
Murray Rubber, Houston

Utah
Microdot/Polyseal, Salt Lake City

Virginia
Century Fasteners Corporation, Richmond
*Cardinal Rubber & Seal Incorporated, Roanoke
Dowty Corporation, Sterling

Table 8. (continued)

Wisconsin
  Trostel, Albert Packings, Lake Geneva
  Grover/Universal Seal, Milwaukee
  Wisconsin Gasket & Mfg. Company, Milwaukee
  Farnam, F. D. Gasket Systems, Nedah

Canada
  Allmetal Screw Products of Canada Ltd., Toronto, Ontario

[a]An asterisk indicates that the company is a product manufacturer.

2

General Design Method

Designers should utilize a systematic method for solving sealing problems. The best method is one the designer has developed personally, has gained confidence in by repeated use, and is versatile enough to be applied to a large variety of seal problems. This chapter presents a general design method that can be modified by designers to meet individual needs. The chapter is divided into three sections, proceeding from general considerations to the more specific. Section 1 presents the general aspects of a useful design method involving gland application, O-ring size, gland dimensions, diametral clearance, and O-ring material. Section II goes into greater detail, discussing such parameters as temperature variations, differential pressure, swell and shrinkage, and so on, and ending with a quick reference table that summarizes the most important design criteria. Section III presents helpful performance charts for the elastomeric materials used for O-ring seals. These charts arrange the elastomeric materials in order of best resistance to temperature, environmental, and physical criteria. This section also ends with a quick summary reference table to aid designers in trade-off studies.

The information presented in this chapter is applicable to both static and dynamic seal problems and should be consulted in addition to information presented in Part II, Specific Applications of Ring Seals.

I. DESIGN METHOD

The following method is a practical approach to designing an O-ring gland and selecting the appropriate seal. Although this

method is presented in the usual design sequence, designers may
begin at any step.

  1. Gland Application and Possible Alternatives: To design an
effective O-ring seal, one must first determine if the parts to be
sealed can be assembled and the seal itself can be installed without difficulty or damage. Much of the responsibility for proper
assembly falls on the designer as he or she provides a safe route
for the O-ring on its way to the groove. Choosing the incorrect
type of gland often leads to assembly problems and sometimes
functional in-service problems.

  For example, consider a piston containing the conventional
male gland and O-ring seal which must slide past a valving port
upon assembly. To prevent the seal from being cut, the port
must be either chamfered or undercut, as shown in Fig. 3. But
the preferred design is to incorporate a female gland in the bore
of the cylinder (a rod seal) so the O-ring seal does not have to
pass over the port, and chamfer the leading edge of the piston,
as in Fig. 4. Additional considerations are presented in Sec. II.

  2. O-Ring Size and Gland Dimensions: The dimensions of an
O-ring and gland are directly dependent on the specific applica-

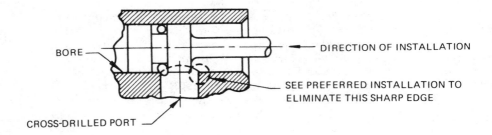

PREFERRED INSTALLATION

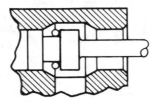

UNDERCUT BORE AS INDICATED

Figure 3. Methods to avoid sharp installation corners

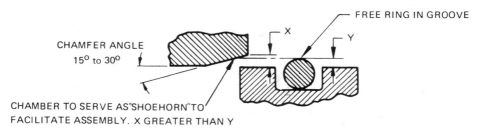

CHAMFER ANGLE
15° to 30°

FREE RING IN GROOVE

X

Y

CHAMBER TO SERVE AS "SHOEHORN" TO
FACILITATE ASSEMBLY. X GREATER THAN Y

Figure 4.   Chamfer piston for rod gland

tion. Static glands are dimensioned to provide the O-ring seal
with an ideal 25% squeeze. The glands for reciprocating seals
are dimensioned for minimum O-ring squeeze to minimize static and
dynamic friction. Rotary seal applications incorporate female glands
dimensioned to provide peripheral compression for the O-ring seal.
Each of these cases is discussed fully under the specific application of static and dynamic seals (Chaps. 4 to 6). Such physical
parameters as O-ring size and the temperature and pressure incurred by the O-ring will help to determine the gland dimensions;
these parameters are discussed in Sec. II.

   3. *Diametral Clearance*: An O-ring seals the leak path between
two concentric parts (i.e., a piston within a cylinder bore). The
O-ring must block the maximum gap between the two parts without
extruding. The maximum gap is the difference between the diameters of the parts—the diametral clearance between the pistons
and the bore. It is based on the fact that the piston may be
forced to one side of the cylinder bore, producing a gap twice that
of the radial difference. If the piston is held concentric in the
bore by bearings or other means, the radial clearance may be
used as the maximum gap. The differential pressure across the
O-ring seal and the hardness of the O-ring seal limit the maximum
diametral clearance allowable. Figure 5 presents the maximum gap
allowed given the O-ring hardness and the fluid pressure across
the O-ring seal. These extrusion curves are very conservative
for most O-ring applications, since they are based on 100,000
pressure cycles, each cycle increasing from zero to the indicated
pressure at the rate of one cycle per second at a test temperature
of 160°F. Because the extrusion curves are conservative and
machining costs are greater for closer tolerances and smaller-
diameter clearances, the maximum gap is usually recommended for

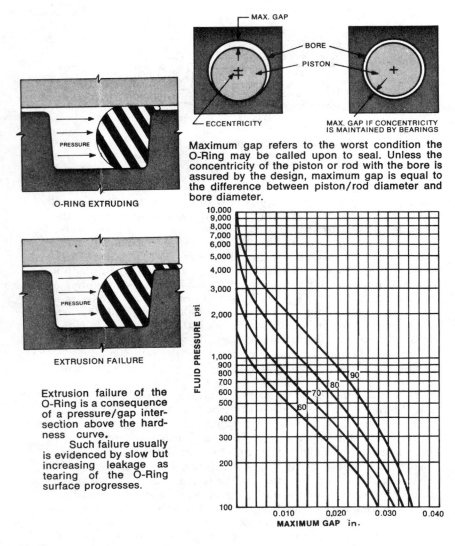

O-RING EXTRUDING

EXTRUSION FAILURE

Extrusion failure of the O-Ring is a consequence of a pressure/gap intersection above the hardness curve.

Such failure usually is evidenced by slow but increasing leakage as tearing of the O-Ring surface progresses.

Maximum gap refers to the worst condition the O-Ring may be called upon to seal. Unless the concentricity of the piston or rod with the bore is assured by the design, maximum gap is equal to the difference between piston/rod diameter and bore diameter.

Figure 5. Maximum gap allowable without extrusion failure (From *The National O-Ring Engineering Manual*, Seven Steps to O-Ring Design, National O-Ring Division, Federal—Mogul Corporation, Downey, California, 1975.)

Table 9. Fluid Compatibility

Fluid	National Elastomer (In order of Recommendation)	Fluid	National Elastomer (In order of Recommendation)	Fluid	National Elastomer (In order of Recommendation)
Acetaldehyde	E	Anderol L-774	B, V, F	Butter (Food)	C
Acetamide	B, E, N	Aniline	E	Butyl Acetate	E
Acetic Acid (Glacial)	E	Aniline Dyes	E	Butyl Acrylate	K
Acetic Acid (30%)	E	Aniline Hydrochloride	E	Butyl Alcohol	B, N, V
Acetic Anhydride	N, E	Animal Oil (Lard)	B, V, F	Butyl Amine	S, E
Acetone	E	Aqua Regia	E, F	Butyl Benzoate	E, V
Acetophenone	E	Aroclor	V	Butyl Carbitol	E
Acetyl Acetone	E	Arsenic Acid	F, E	Butylene	V, F, B
Acetyl Chloride	V, F	Askarel	V, B, F	Butyl Ether	K
Acetylene	E, B	Asphalt	V, K	Butyl Butyrate	E, V
Acetylene Tetrabromide	V, E			Butyl Oleate	V, E
Aerozine 50	E	Barium Chloride	B, E	Butyl Stearate	V, F, B
Air (Below 300 °F.)	E, B	Barium Hydroxide	B, E	Butyraldehyde	E, K
Air (Above 300 °V.)	S, V, F	Barium Sulfide	B, E	Butyric Acid	V, E
Alkazene	V, F, K	Beer (Food)	C	Calcine Liquors	B, E
Alum	B, E	Beet Sugar Liquors (Food)	C	Calcium Acetate	E
Aluminum Acetate	E, B	Benzaldehyde	E	Calcium Bisulfite	B, N, V
Aluminum Bromide	B, E, N	Benzene	V, F, K	Calcium Chloride	B, E, N
Aluminum Chloride	B, E, N	Benzenesulfonic Acid	V, F, N	Clacium Hydroxide	B, E, N
Aluminum Fluoride	B, E, N	Benzine	V, F, K	Calcium Hypochlorite	E, V
Aluminum Nitrate	B, E, N	Benzoic Acid	V, F, K	Calcium Nitrate	B, E, N
Aluminum Sulfate	B, E, N	Benzochloride	V, F, E	Calcium Sulfide	B, E, N
Amines	E	Benzophenone	V, F, E	Cane Sugar Liquors (Food)	C
Ammonia (Anhydrous)	E, B	Benzyl Alcohol	V, F, E	Carbitol	E, B
Ammonia (Liquid)	E, B	Benzyl Benzoate	V, F, E	Carbolic Acid	E, F
Ammonium Carbonate	E, N	Benzyl Chloride	V, F	Carbon Bisulfide	V, F
Ammonium Chloride	B, E, N	Black Sulfate Liquors	E	Carbonic Acid	E, N
Ammonium Hydroxide	E, B	Blast Furnace Gas	V, S	Carbon Dioxide	B, E
Ammonium Nitrate	B, E	Bleach Liquor	E, V	Carbon Disulfide	V, F
Ammonium Nitrite	B, E	Borax	B, E	Carbon Monoxide	B, E
Ammonium Persulfate	E	Bordeaux Mixture	E, V	Carbon Tetrachloride	V, F
Ammonium Phosphate	B, E	Boric Acid	B, E	Castor Oil (Food)	C
Ammonium Sulfate	B, E	Boron Fluids (HEF)	V, F	Cellosolve	E, K
Ammonium Sulfide	B, E	Brake Fluid (Automotive)	E	Cellosolve Acetate	E, K
Amyl Acetate	E	Bromine	E, V, F	Cellulubes	E, K
Amyl Alcohol	E	Bromine Water	E, F	China Wood Oil	B, V
Amyl Borate	B, N, K	Bromobenzene	V, F	Chlorinated Solvents	V, F
Amyl Chloronaphthalene	V, K	Bunker Oil	B, V, F	Chlorine	E, F
Amyl Chloride	V, F	Butadiene	V, F, E	Chlorine Dioxide	E
Amyl Naphthalene	V, F	Butane	B, V, K	Chloroacetic Acid	E

KEY:

B, C	Nitrile	J	Butyl	V	Fluorocarbon	G	Styrene-butadiene
E	Ethylene-Propylene	K	Polysulfide	S	Silicone	L	Polyacrylate
N	Neoprene	H	Hypalon*	F	Fluorosilicone	U	Polyurethane

Table 9 (Continued)

Fluid	National Elastomer (In order of Recommendation)	Fluid	National Elastomer (In order of Recommendation)	Fluid	National Elastomer (In order of Recommendation)
Chloroacetone	E	Dichlorobutane	V, B	Ethyl Formate	V, F, E
Chlorobenzene	V, F	Dichloro-Isopropyl Ether	K	Ethyl Hexanol	B, E
Chlorobromomethane	V, F, E	Dicyclohexylamine	B	Ethyl Mercaptan	V
Chlorobutadiene	V, F, E	Diesel Oil	B, V, F	Ethyl Oxalate	V, K, F
Chlorododecane	V, F, E	Diester Syn. Lubricants	B, V, F	Ethyl Pentachlorobenzene	V, F
Chloroform	V, F	Diethylamine	E	Ethyl Silicate	E, B
Chloronaphthalene	V, F	Diethyl Ether	K	Ferric Chloride	E, B
Chlorotoluene	V, F	Diethylene Glycol	E, B	Ferric Nitrate	E, B
Chlorox	E, F	Diethyl Sebacate	V, E	Ferric Sulfate	E, B
Chlorophenol	V	Difluorodibromomethane	E	Fluoboric Acid	E, N
Chrome Plating Solutions	E	Diisobutylene	V, B, K	Fluorolube	E, B
Chromic Acid	E	Diisooctyl Sebacate	V, E	Fluorochloroethylene	V
Citric Acid (Food)	C	Diisopropyl Ketone	E	Formaldehyde	E, B
Cobalt Chloride	B, E	Dimethyl Formamide	B, S, E	Formic Acid	E, N
Coconut Oil (Food)	C	Dimethyl Phthalate	E, V	Freon 11	K, V, B
Cod Liver Oil (Food)	C	Dioctyl Phthalate	E, V	Freon 12	N, B, K
Coke Oven Gas	V, F, S	Dioctyl Sebacate	V, E	Freon 13	N, B, K
Coolanol	N, V, F	Dioxane	E	Freon 13B1	N, B, K
Compass Fluid	B, E	Dioxolane	E	Freon 14	N, B, K
Copper Acetate	E	Dipentene	V, K, B	Freon 21	N
Copper Chloride	B, E	Diphenyl	V, F, K	Freon 22	N, K, E
Copper Cyanide	B, E	Diphenyl Oxides	V, F	Freon 31	N, E
Copper Sulfate	B, E	Dowtherm A or E	V, F	Freon 32	N, E
Corn Oil (Food)	C	Dry Cleaning Fluids	V, F	Freon 112	K, B
Cottonseed Oil (Food)	C			Freon 113	N, B, K
Creosote	B, V, F	Epichlorohydrin	E	Freon 114	N, B, E
Cresols	F, V	Ethanolamine	N, E, B	Freon 114B2	K, N
Crude Oil	V, F	Ethers	K	Freon 115	B, N, E
Cutting Oil	B, V, F	Ethyl Acetate	E, K	Freon 142b	N, B, E
Cyclohexane	B, V, F	Ethyl Acetoacetate	E, K	Freon 152a	N, B, E
		Ethyl Acrylate	E, K	Freon 218	N, B, E
		Ethyl Alcohol	E, B	Freon C316	N, B, K
Decalin	V, F, K	Ethyl Benzene	V, F, E	Freon C318	N, B, E
Decane	B, V, F	Ethyl Benzoate	V, F, K	Freon BF	K, B, N
Deionized Water	E, B	Ethyl Cellosolve	E, K	Freon MF	K, V, B
Denatured Alcohol	E, B	Ethyl Cellulose	B, N, E	Freon TF	N, B, E
Detergents	E, B	Ethyl Chloride	E, B, N	Fuel Oil	B, V, F
Developing Fluids (Photo)	E, N	Ethyl Chlorocarbonate	V, F	Fumaric Acid	B, V, F
Diacetone	E	Ethyl Chloroformate	V, F	Furfural	E
Diacetone Alcohol	E	Ethylene Chloride	V	Furfuryl Alcohol	E
Dibenzyl Ether	K, E	Ethylene Chlorohydrin	V, E		
Dibenzyl Sebacate	V, E	Ethylene Diamine	E, B	Gallic Acid	V, F, E
Dibromoethyl Benzene	V, F	Ethylene Dibromide	V	Gasoline (Automotive)	B, V, F
Dibutylamine	E, N	Ethylene Dichloride	V	Gelatin (Food)	C
Dibutyl Ether	K	Ethylene Glycol	E, B	Glucose (Food)	C
Dibutyl Phthalate	E, K	Ethylene Oxide	E	Glue	B, E
Dibutyl Sebacate	E, K	Ethylene Trichloride	V	Glycerine	B, E
Dichlorobenzene	V, K	Ethyl Ether	K	Glycols	E, B

Table 9 (Continued)

Fluid	National Elastomer (In order of Recommendation)	Fluid	National Elastomer (In order of Recommendation)	Fluid	National Elastomer (In order of Recommendation)
Green Sulfate Liquors	E	Lead Nitrate	E, B	Monovinyl Acetylene	E, B
		Lead Sulfamate	N, E, V		
HEF-2	V	Liqroin	B, V, F		
Helium	E	Lime Bleach	B, E, V	Naphtha	V, B, F
Heptane	B, V, F	Lime Sulfur	E, V	Naphthalene	V, F, K
Hexaldehyde	E, N	Lindol	E, V	Napthenic Acid	V, F, B
Hexane	B, V, F	Linoleic Acid	S, N	Natural Gas	B, V, E
Hexene	V, F, K	Linseed Oil	B, V, F	Neatsfoot Oil	B, V, F
Hexyl Alcohol	B, V, F	Liquid Oxygen	S, V	Nickel Acetate	E, B
Houghto-Safe 271	B, E, V	Liquefied Petroleum Gas		Nickel Chloride	E, B
620	B, E, V	(LPG)	B, V, K	Nickel Sulfate	E, B
1010	E, V	Lubricating Oils	B, V, F	Nitric Acid (Dilute)	E
1055	E, V	Lye	E	Nitrobenzene	V
1120	E, V			Nitroethane	N, E
5040	B, V, F	Magnesium Chloride	E, B	Nitrogen	E, B
Hydrolube	B, E, V	Magnesium Hydroxide	E, V	Nitromethane	K, E
Hydraulic Oil (Petroleum)	B, V, F	Magnesium Sulfate	E, B,	Nitropropane	K, E
Hydrazine	E	Magnesium Sulfite	E, B	Non-Toxic Compound	
Hydrobromic Acid	E	Maleic Acid	V, K	(Food)	C
Hydrochloric Acid	E	Maleic Anhydride	V		
Hydrocyanic Acid	E	Malic Acid	B, V, F	Octadecane	B, V, F
Hydrofluoric Acid	E	Mercuric Chloride	E, B	Octane	B, V, K
Hydrofluosilicic Acid	E	Mercury	E, B	Octyl Alcohol	E, V
Hydrogen	E	Mesityl Oxide	E, K	Oleic Acid	B
Hydrogen Peroxide	F, V, E	Methyl Acetate	E, K	Oleum Spirits (Food)	C
Hydrogen Sulfide	E, B	Methyl Acrylate	E, K	Oleum	E
Hydroquinone	V, F	Methylacrylic Acid	E, N	Olive Oil (Food)	C
Hypochlorous Acid	E	Methyl Alcohol	E, N	Oronite 8200	N, B, V
		Methyl Bromide	V, F	Oronite 8515	N, B, V
Iodine	V, E	Methyl Cellosolve	E	Ortho-Dichlorobenzene	V, F
Isobutyl Alcohol	E, B	Methyl Chloride	V, F, K	OS-45	N, V, F
Isobutyl Butyrate	E, B	Methyl Cyclopentane	V, F, K	Oxalic Acid	E, V
Isododecane	B, V, F	Methylene Chloride	V, F	Oxygen (Gaseous)	S, E
Iso-Octane	B, V, F	Methylene Dichloride	V, F	Ozone	E, N
Isophorone	E	Methyl Ether	E, B		
Isoproptyl Acetate	E, K	Methyl Ethyl Ketone	E, K		
Isopropyl Alcohol	E, B	Methyl Formate	N, E	Paint Solvents	K
Isopropyl Chloride	V, F	Methyl Isobutyl Ketone	E, K	Palmitic Acid	B, V, F, K
Isopropyl Ether	B, K, N	Methyl Isopropyl Ketone	E, K	Para-Dichlorobenzene	V, F
		Methyl Methacrylate	K	Peanut Oil (Food)	C
JP-1 Thru JP-6 Fuel	B, V, F	Methyl Oleate	V, E	Pentane	B, V
		Methyl Salicylate	E	Perchloric Acid	F, E
Kerosene	B, V, F	Milk (Food)	C	Perchlorethylene	V, K, F
		Mineral Oil (Food)	C	Petrolatum	B, V, F
Lacquers	K, E	Monomethylaniline	V	Petroleum Oils	B, V, F
Lactic Acid (Food)	C	Monobromobenzene	V, K	Phenol	F, V
Lard (Food)	C	Monochlorobenzene	V, F	Phenylbenzene	V, F, K
Lead Acetate	E	Monoethanolamine	E	Phenylethyl Ether	K

Table 9   (Continued)

Fluid	National Elastomer (In order of Recommendation)	Fluid	National Elastomer (In order of Recommendation)	Fluid	National Elastomer (In order of Recommendation)
Phenylhydrazine	V, E	Silicone Greases	E, B	Tertiary Butyl Alcohol	V, B, E
Phorone	E	Silicone Oils	E, B	Tertiary Butyl Catechol	V, E
Phosphate Esters, Alkyl	E	Siver Cyanide	E, B	Tertiary Butyl Mercaptan	V
Phosphate Esters, Aryl	V, E	Silver Nitrate	E, B	Tetrabromoethane	V, F
Phosphoric Acid (45%)	E	Skydrol	E	Tetrachloroethane	V, F
Phosphorous Trichloride	E, V	Soap Solutions	E, B	Tetrachloroethylene	V, F
Pickling Solution	E	Sodium Acetate	E, B	Tetraethyl Lead	V, F, B
Picric Acid	E	Sodium Bicarbonate	E, B	Tetrahydrofuran	E, K
Pinene	V, F, B	Sodium Borate	E, B	Tetralin	V, F
Pine Oil	B, V, F	Sodium Bisulfate	E, B	Titanium Tetrachloride	V, F
Plating Solutions	E	Sodium Bisulfite	E, B	Toluene (Toluol)	V, F
Pneumatic Service	B, E, N	Sodium Carbonate	E, B	Transformer Oil	B, V, F
Polyvinyl Acetate	E	Sodium Chloride	E, B	Triacetin	E
Potassium Acetate	E	Sodium Cyanide	E, B	Tributoxyethyl Phosphate	E, V
Potassium Chloride	E, B	Sodium Dichromate	E, B	Tributyl Mercaptan	V, E
Potassium Cyanide	E, B	Sodium Hydroxide	E, B	Tributyl Phosphate	E, K
Potassium Dichromate	E, B	Sodium Hypochlorite	E, N	Trichloroethane	V, F
Potassium Hydroxide	E	Sodium Metaphosphate	E, B	Trichloroacetic Acid	E, B
Potassium Nitrate	E, B	Sodium Nitrate	E, B	Trichloroethylene	V, F
Potassium Sulfate	E, B	Sodium Perborate	E, B	Tricresyl Phosphate	E
Potassium Sulfite	E, B	Sodium Peroxide	E, V	Triethanolamine	E
Prestone	E, B	Sodium Phosphate	E, B	Trinitrotoluene	V, N
Propane	B, V	Sodium Silicate	E, B	Trioctyl Phosphate	E
Propyl Acetate	E, K	Sodium Sulfate	E, B	Trisodium Phosphate	E, B
Propyl Acetone	E, K	Sodium Sulfide	E, B	Tung Oil	B, V, F
Propyl Alcohol	E	Sodium Sulfite	E, B	Turbine Oil	V, B
Propyl Nitrate	E	Sodium Thiosulfate	E, B	Turpentine	B, V, F
Propylene	V, F, K	Soybean Oil (Food)	C		
Propylene Oxide	E	Stannic Chloride	E, B	Unsym. Dimethyl Hydrazine	E
Pyranol	B, V, F	Stannous Chloride	E, B		
Pydraul 150	E, V	Steam	E, B	Varnish	V, K, F
A-200	V, F, K	Stearic Acid	B, E	Vegetable Oil (Food)	C
A C	E, V	Stoddard Solvent	B, V, F	Versilube F-50	E, B
F-9	E, V	Styrene	V, F	Vinegar (Food)	C
625	E, V	Sucrose Solutions (Food)	C		
Pyridine Oil	E	Sulfur	N, E	Water (Food)	C
Pyrolube	V, E	Sulfur Chloride	V, F	Whiskey (Food)	C
		Sulfur Dioxide	E, V	Wine (Food)	C
Red Oil (MIL-H-5606)	B, V, F	Sulfur Hexafluoride	N, E	White Pine Oil	V, F, B
RJ-1	B, V, F	Sulfur Free Compound	N		
RP-1	B, V, F	Sulfur Trioxide	V, E	Xylene (Xylol)	V, F, B
Rapeseed Oil	E, V	Sulfuric Acid	E	Xylidenes	B, E
		Sulfurous Acid	E		
Sal Ammoniac	E, B			Zinc Acetate	E, B
Salicylic Acid	E, V			Zinc Chloride	E, B
Salt Water	E, B	Tannic Acid	E, B	Zinc Sulfate	E, B
Sewage	E, B	Tar	V, B		
Silicate Esters	N, V, F	Tartaric Acid	B, V, F		

Source: *Seven Steps to O-Ring Design*, National, Federal-Mogul
The National O-Ring Engineering Manual, Seven Steps
to O-Ring Design, National O-Ring Division, Federal—
Mogul Corporation, Downey, California, 1975.

the O-ring hardness and pressure incurred. The 70-durometer-hardness O-ring is the most common for elastomeric materials, although some materials, such as ethylene propylene, are more commonly compounded at 80 durometer by some manufacturers. The 90-durometer curve is equivalent to a 70-durometer O-ring with backup rings. The designer should also realize that the durometer tolerance is usually ±5 points for common O-ring seal materials.

4. *O-Ring Material*: Selecting the correct O-ring material is usually a compromise between chemical compatibility and physical capability. The O-ring material must not be attacked by the fluid it seals and it must not be easily abraded, scored, permanently compressed, or otherwise physically damaged by the environment. Table 6 should be consulted for the physical capabilities and comparative properties of O-ring materials, and Table 9 should be consulted for chemical compatibility.

II. SPECIFIC DESIGN CONSIDERATIONS

The designer must consider a number of specific design parameters relative to the particular application of the O-ring seal required. Such parameters as O-ring size, O-ring stretch, temperature variations, differential pressure, swell and shrinkage, corrosion, and radiation must be considered as applications become more complex and sophisticated. These parameters are discussed in this section and a summary table listing the important criteria is presented for the designer's quick reference.

A. O-Ring Size

The dimensions of elastomeric O-rings vary because of mold shrinkage during production. Tolerances of O-ring cross section and internal diameter are given in AS568. These tolerances must be taken into account when calculating minimum and maximum O-ring squeeze and minimum and maximum gland dimensions. Most manufacturers' gland design tables provide for these tolerances.

B. O-Ring Stretch

When installing an O-ring in a groove, the inside diameter (ID) should not be stretched greater than 100%, except for very small diameter O-rings. Sufficient time should be allowed for the O-ring to return to its original diameter before final assembly of parts. When installed in the groove, the ID of the O-ring should not ex-

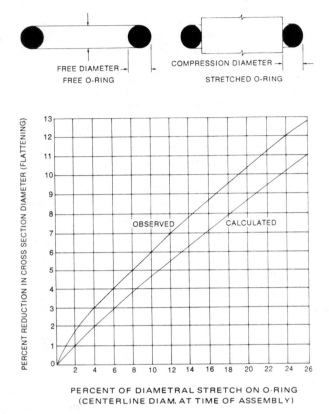

Figure 6. Loss of compression diameter due to stretch (From *O-Ring Handbook OR5700*, Parker Seal Co., Lexington, Ky., January 1977)

ceed 5% stretch. Most elastomeric materials deteriorate under excessive stretch conditions. Ethylene propylene, fluorocarbon, polyurethane, and neoprene are best for high-stretch applications, while nitrile rubber is poorest [2, p. A4-6].

The cross section of an O-ring is reduced when its ID is stretched. (Conversely, the cross section is increased when the O-ring is put into peripheral compression. This is discussed in Chap. 6.) This reduction in cross section is presented in Fig. 6.

When the diametral stretch is greater than 2 or 3%, the O-ring cross section is observed to be reduced by 1.8 or 2.5%, and therefore the depth of the gland into which the O-ring is to be installed must be reduced to maintain the necessary squeeze on the O-ring. The percent reduction in cross-section diameter is approximately one-half the percent of diametral stretch on the O-ring. Examples using this relationship are presented in Chap. 4.

C. Temperature Variation

An O-ring application involving a temperature variation of several hundred degrees will experience dimensional differences that result in changing O-ring squeeze and sealing capability. Because the coefficient of thermal expansion for elastomers (80 to 150 X 10^{-6} in./in. °F) is approximately 10 times that of metals (6 to 13 X 10^{-6} in./in. °F), it is necessary to provide expansion volume for the O-ring when designing the gland. If insufficient expansion volume is not provided, the O-ring either will extrude or, if trapped, may fracture the gland. At the other extreme, cryogenic temperatures will cause O-ring shrinkage, eventual loss of squeeze, and leakage. The design tables in Chaps. 3, 4, and 6 were devised such that proper O-ring sealing is ensured for the temperature ranges listed for the standard elastomeric materials (Fig. 2). However, the designer may want to modify the gland dimensions for an application that experiences only high temperature or low temperature in order to maintain a particular squeeze on the O-ring. The following coefficients of thermal expansion will aid the designer:

Material	Change (in./in. °F)
Nitrile	6.2×10^{-5}
Neoprene	7.6×10^{-5}
Fluorocarbon	9.0×10^{-5}
Ethylene propylene	8.9×10^{-5}
Other organic elastomers	$8-12 \times 10^{-5}$
Silicones	$12-15 \times 10^{-5}$

An example in the use of the coefficient of thermal expansion as applied to gland design is presented in Design Example 3 (Chap. 4).

D. Differential Pressure

Differential pressure is the difference between the pressure acting
on one side of an O-ring cross section and the pressure acting
on the other side of the O-ring cross section. In general, the
differential pressure must be over 50 psi to distort the O-ring
beyond its initial installation squeeze. Below this differential
pressure, as in a partial vacuum, the squeeze and resilience of
the O-ring maintains the seal. At differential pressures greater
than 1500 psi the designer has to consider methods to prevent excess O-ring distortion and extrusion. Maximum diametral clear-

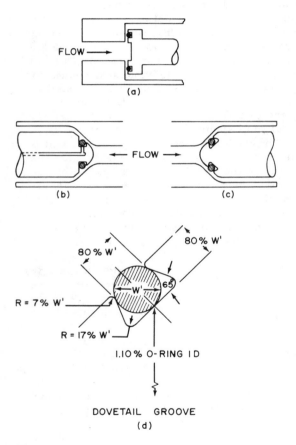

Figure 7. Blowout prevention

ance or gap may be reduced as shown in Fig. 6, backup rings may be used, or a harder-durometer material may be chosen. Elastomers are usually considered to be incompressible, but the designer may have occasion to use the general compression modulus of 500,000 lb/in.^{3} [1, p. 6].

Differential pressure across an O-ring cross section causes two additional phenomena in dynamic seal applications. Besides trying to extrude the O-ring out of the gap, differential pressure increases O-ring friction and is responsible for O-ring "blowout" in dynamic seal applications. O-ring blowout usually occurs when the differential pressure is great enough to hold the O-ring against the face of a face seal while the grooved portion of the seal moves away. The net result is the O-ring is blown out of the groove and the face seal cannot reseal. This may occur in straight face-seal design, as shown in Fig. 7a. The design shown in Fig. 7b incorporates a central hole that allows the pressure to equalize around the O-ring as the valve opens. The dovetail goove shown in Fig. 7c captures the O-ring within the groove. The configuration of the dovetail groove, as specified in Fig. 7d, is expensive to machine and should be used only when absolutely necessary. The cross-sectional diameter W' is as installed and includes temperature and fluid swell considerations.

E. Swell and Shrinkage

Swell and skrinkage refer to the percentage change in O-ring volume caused by the O-ring absorbing or reacting with the fluid being sealed. O-ring swell may enhance sealing effectiveness, although it is usually accompanied by a decrease in hardness which may promote O-ring extrusion under high differential pressure. As a rule of thumb, Parker Seal Company allows up to 50% swell for static O-ring applications and up to 20% swell for dynamic applications [2, p. A3-7]. This author usually designs for and recommends only 10% maximum swell for dynamic applications. O-ring shrinkage always decreases sealing capability. Shrinkage usually causes increase in hardness and compression set; the net effect being a smaller, harder, and less resilient O-ring cross section. Because shrinkage also involves negative hardness and compression set effects, the percentage shrinkage must be well below the initial percentage squeeze of the installed O-ring cross section. No more than 3% shrinkage is the recommended allowance for dynamic O-ring applications.

Glands may be designed to accommodate O-ring swell or shrinkage. Most O-ring seals will swell slightly even in compatible fluids,

but in cases requiring material compromise, glands and grooves
may be enlarged to provide for greatly swollen seals. Conversely,
glands and grooves may be reduced in cases where the seal will
be known to shrink in operation. This may make assembly of the
seal more difficult, but the O-ring may be soaked in the fluid to
preshrink it before installation.

F. Corrosion

Corrosion between the fluid and the gland material must be
considered when designing an effective seal. Corrosion of the
metal surfaces of the gland may cause pitting and fretting to the
extent of virtually changing the surface finish—promoting leakage
across the seal. Electrolytic and galvanic effects should always
be considered when designing a system incorporating O-ring
seals. Dissimilar materials such as an aluminum piston in a steel
cylinder involving an electrolytic fluid such as seawater will
guarantee corrosion. Pitting corrosion will usually start at mini-
mum-clearance surfaces that trap oxygen, as under O-ring seals.
It is therefore important to use the same material or equivalents
throughout a system, machine surfaces to smooth finishes,
throughly clean these surfaces before assembly, and maintain them
during operation. Nonoxidizing lubricants are recommended in
both static and dynamic seal applications, but caution must be
taken to ensure that the lubricant itself does not promote corrosion. In the case of the seawater hydraulic cylinder, it is best
to use different types of stainless steel materials for the piston
and cylinder, and ethylene propylene or nitrile O-ring seals lubricated with a noncorrosive silicone grease.

G. Radiation

Effects of gamma-wave and neutron-particle radiations must be
considered when selecting the proper O-ring material for applications involving radiation. Gamma radiation is the most severe
form of radiation, affecting O-ring hardness, tensile strength,
compression set, and low-temperature properties. According to
the O-ring literature, most elastomers will begin to experience
physical and chemical damage between 10^8 and 10^{10} ergs/g of
gamma radiation. Nuclear reactors have usually incorporated
ethylene propylene and fluorocarbon O-ring materials, not because
these materials have any exceptional resistance to radiation, but
because of their resistance to water and high temperatures.

Table 10. Designer's Quick Reference: Specific Design Considerations[a]

Design consideration	Important criteria
O-ring stretch	ID stretch no greater than 100% when installing O-ring ID stretch no greater than 5% when installed in groove Percent reduction in cross-section diameter is approximately one-half the percent diametral stretch
Temperature variation	Coefficient of thermal expansion for elastomers is between 60 and 150 X 10^{-6} in./in. °F (10 times that of metals)
Differential pressure	Normal O-ring, use 50 to 1500 psi Reduction in diametral clearance, harder-durometer material, or backup rings recommended for pressures greater than 1500 psi General compression modulus is 500,000 lb/in.^{3}
Swell and shrinkage	Static applications: 50% maximum allowable swell 3% maximum allowable shrinkage Dynamic applications: 10% maximum allowable swell 3% maximum allowable shrinkage
Corrosion	Minimize electrolytic and galvanic effects; correct lubricants deter corrosion
Radiation	Elastomeric deterioration starts at 10^8 erg/g of gamma radiation

[a]This table summarizes the important criteria discussed in Sec. II.

Radiation may be considered a genral deteriorant, increasing the negative effects of other environmental conditions, such as temperature, pressure, abrasion, and chemical reaction, on the elastomeric material. Table 10 summarizes the important criteria discussed in this section.

III. MATERIAL PERFORMANCE OF ELASTOMERS

Any good design method must include function or compatibility trade-offs, depending on the specific O-ring seal application. Most trade-off studies involve a compromise between the various physical capabilities of elastomeric seal materials and their resistance to environmental conditions. For example, ethylene propylene O-rings would make excellent piston-ring seals because of their resistance to physical abuse, but because most pumps use petroleum-base lubricants (in which ethylene propylene dissolves), ethylene propylene O-rings are not normally used for piston seals.

A series of material performance charts have been arranged to aid the designer in trade-off studies leading to the selection of the proper elastomeric material. These charts present temperature, environmental, and physical criteria. The material's relative resistance is rated according to excellent, good, fair, and poor performance, allowing the designer to compare materials at a glance. Chart 1A lists materials in the same order as in Table 6 and in Sec. III of Chap. 1: the most common elastomeric materials first. Chart 1B lists the materials most resistant to temperature deviations first. Therefore, silicone has an excellent temperature rating (weighted equivanlent of 8), flurosilicone performs between excellent and good (weighted equivalent of 7.5), nitrile has a good performance of 6.0, and so on. Chart 1C arranges the materials in order of most resistance to environmental criteria, fluorocarbon's environmental resistance being averaged at 7.6, close to excellent; ethylene propylene is averaged at 6.0, a good rating. Chart 1D lists the materials in order of best physical resistance, ethylene propylene being the best at an average rating across the physical criteria band of 6.7 and nitrile being averaged at 6.0, a good rating. It must be emphasized that these charts are presented as an aid to the designer. The order of arrangement and average weighted equivalent numbers are only indicators of material performance. The designer must always review the specific criteria (e.g., weather resistance, tear resistance, tensile strength, etc.) when making function or compatibility trade-offs.

The average weighted equivalent numbers within each of the three criteria bands—temperature, environmental, and physical—

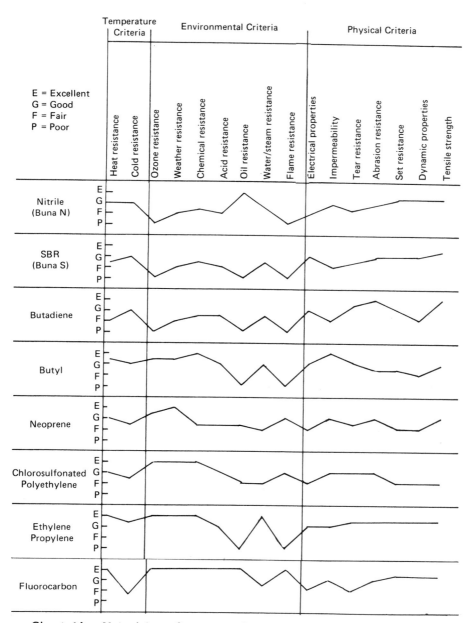

Chart 1A. Material performance of elastomers: most common material first

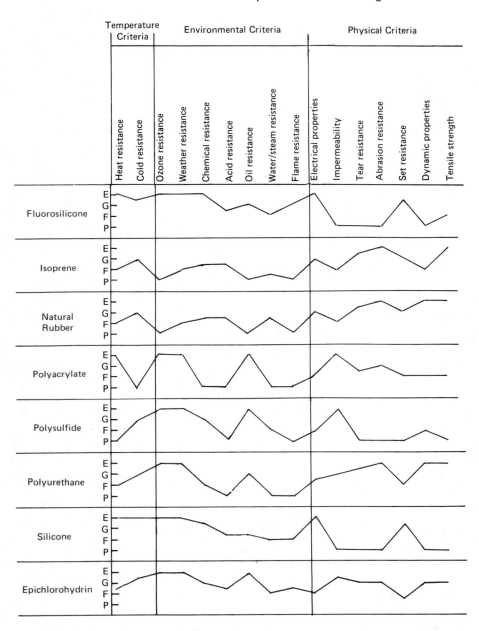

Chart 1A (continued)

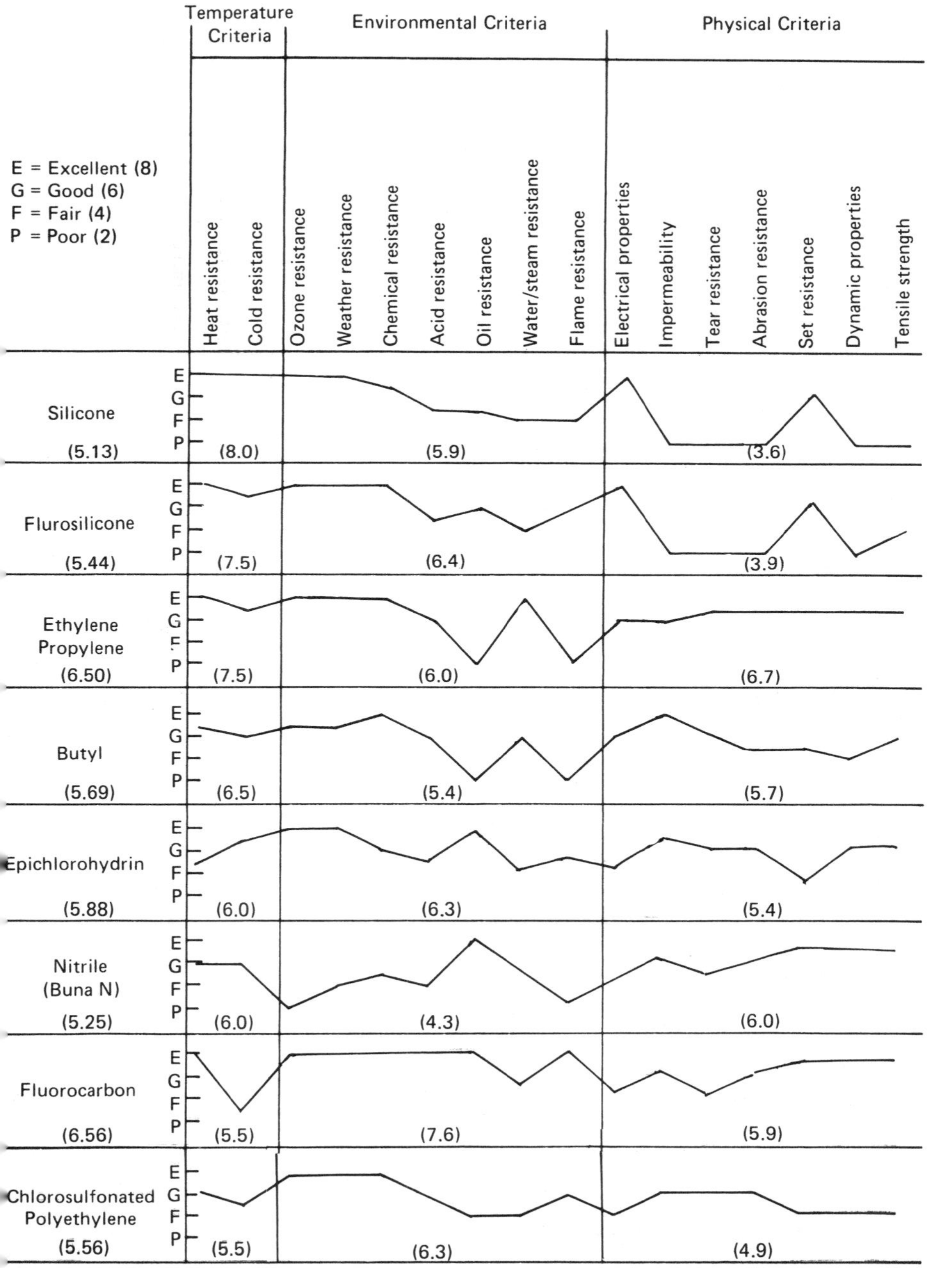

Chart 1B. Elastomers: best resistance to temperature

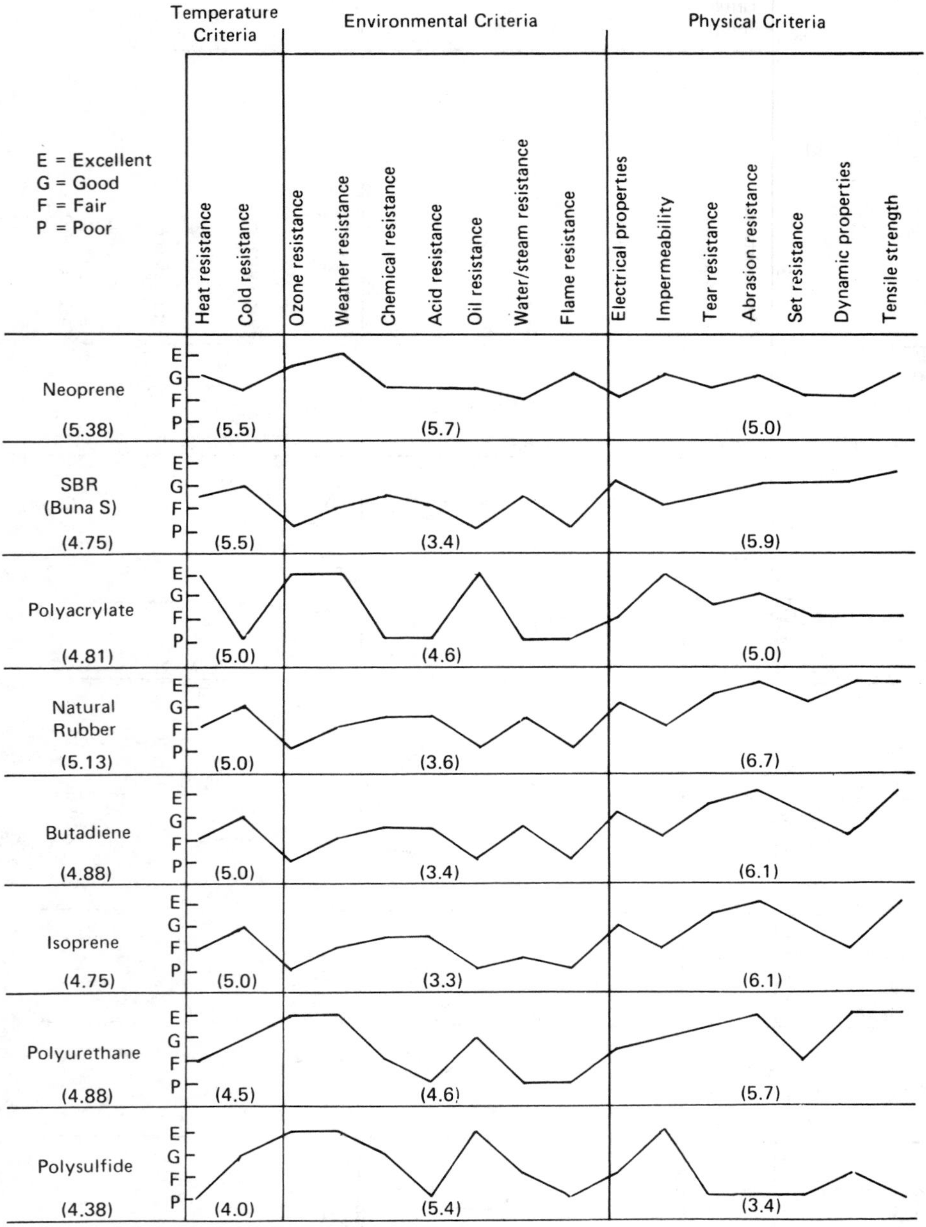

	E = Excellent G = Good F = Fair P = Poor		Temperature Criteria		Environmental Criteria							Physical Criteria						
			Heat resistance	Cold resistance	Ozone resistance	Weather resistance	Chemical resistance	Acid resistance	Oil resistance	Water/steam resistance	Flame resistance	Electrical properties	Impermeability	Tear resistance	Abrasion resistance	Set resistance	Dynamic properties	Tensile strength
Neoprene (5.38)			(5.5)		(5.7)							(5.0)						
SBR (Buna S) (4.75)			(5.5)		(3.4)							(5.9)						
Polyacrylate (4.81)			(5.0)		(4.6)							(5.0)						
Natural Rubber (5.13)			(5.0)		(3.6)							(6.7)						
Butadiene (4.88)			(5.0)		(3.4)							(6.1)						
Isoprene (4.75)			(5.0)		(3.3)							(6.1)						
Polyurethane (4.88)			(4.5)		(4.6)							(5.7)						
Polysulfide (4.38)			(4.0)		(5.4)							(3.4)						

Chart 1B (continued)

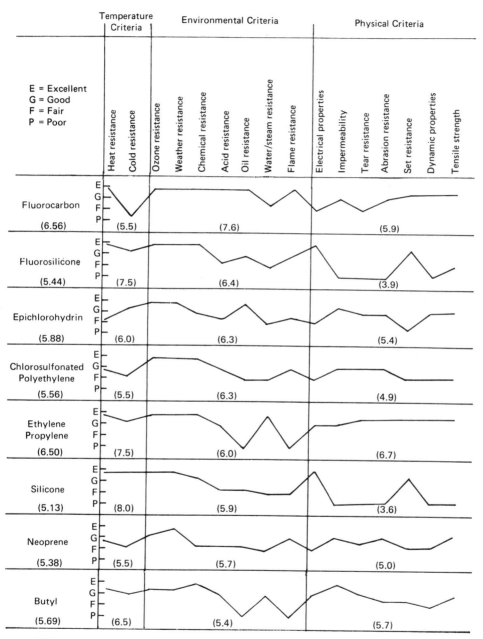

Chart 1C. Elastomers: best environmental resistance

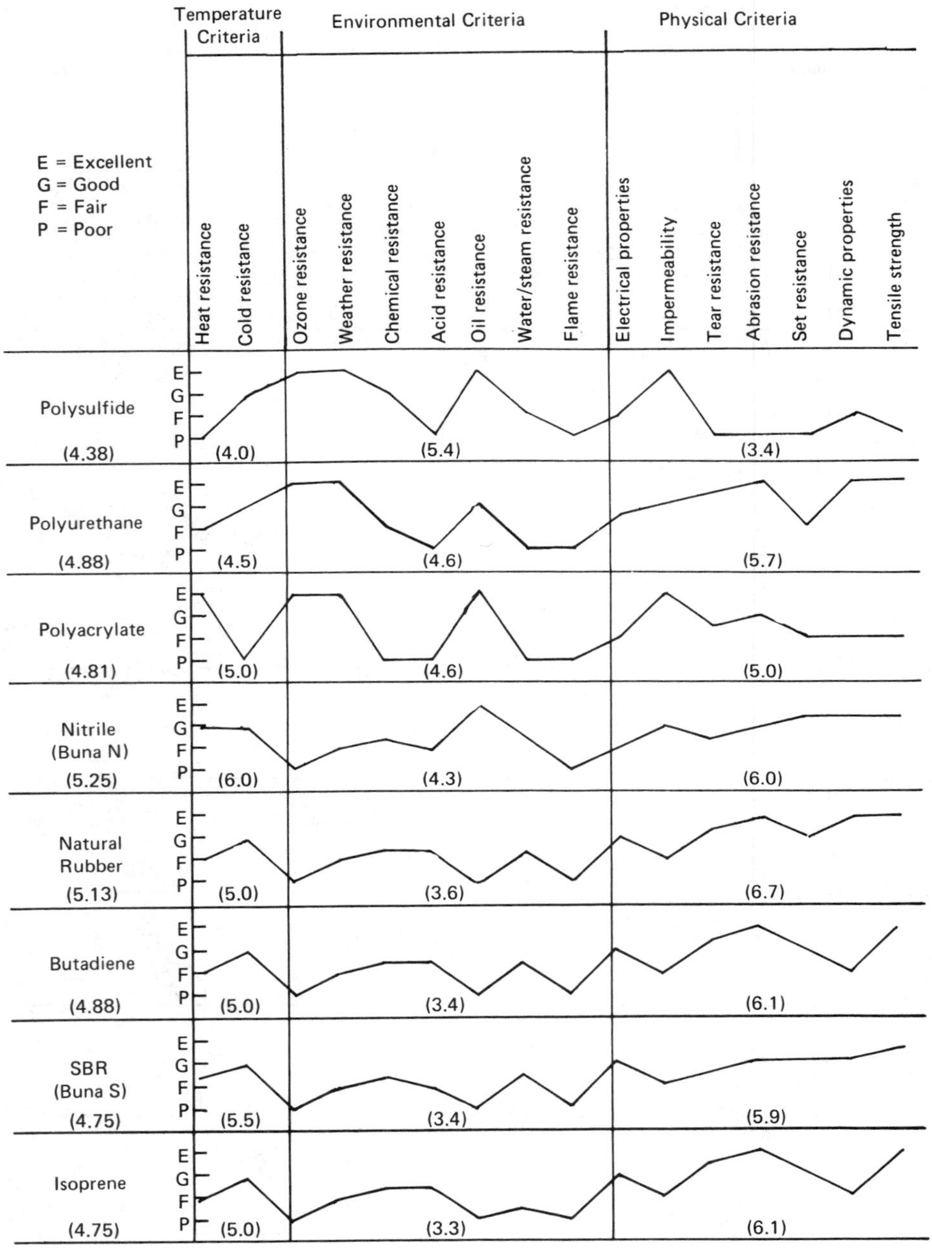

This is a comparison chart showing material property ratings. The data is presented as line graphs for each material across property columns. The following table captures the rating scale, column headers, material names with their overall averages, and the sub-averages printed beneath each criteria group.

Rating scale:
- E = Excellent
- G = Good
- F = Fair
- P = Poor

	Temperature Criteria		Environmental Criteria							Physical Criteria						
Material	Heat resistance	Cold resistance	Ozone resistance	Weather resistance	Chemical resistance	Acid resistance	Oil resistance	Water/steam resistance	Flame resistance	Electrical properties	Impermeability	Tear resistance	Abrasion resistance	Set resistance	Dynamic properties	Tensile strength
Polysulfide (4.38)	(4.0)		(5.4)							(3.4)						
Polyurethane (4.88)	(4.5)		(4.6)							(5.7)						
Polyacrylate (4.81)	(5.0)		(4.6)							(5.0)						
Nitrile (Buna N) (5.25)	(6.0)		(4.3)							(6.0)						
Natural Rubber (5.13)	(5.0)		(3.6)							(6.7)						
Butadiene (4.88)	(5.0)		(3.4)							(6.1)						
SBR (Buna S) (4.75)	(5.5)		(3.4)							(5.9)						
Isoprene (4.75)	(5.0)		(3.3)							(6.1)						

Chart 1C (continued)

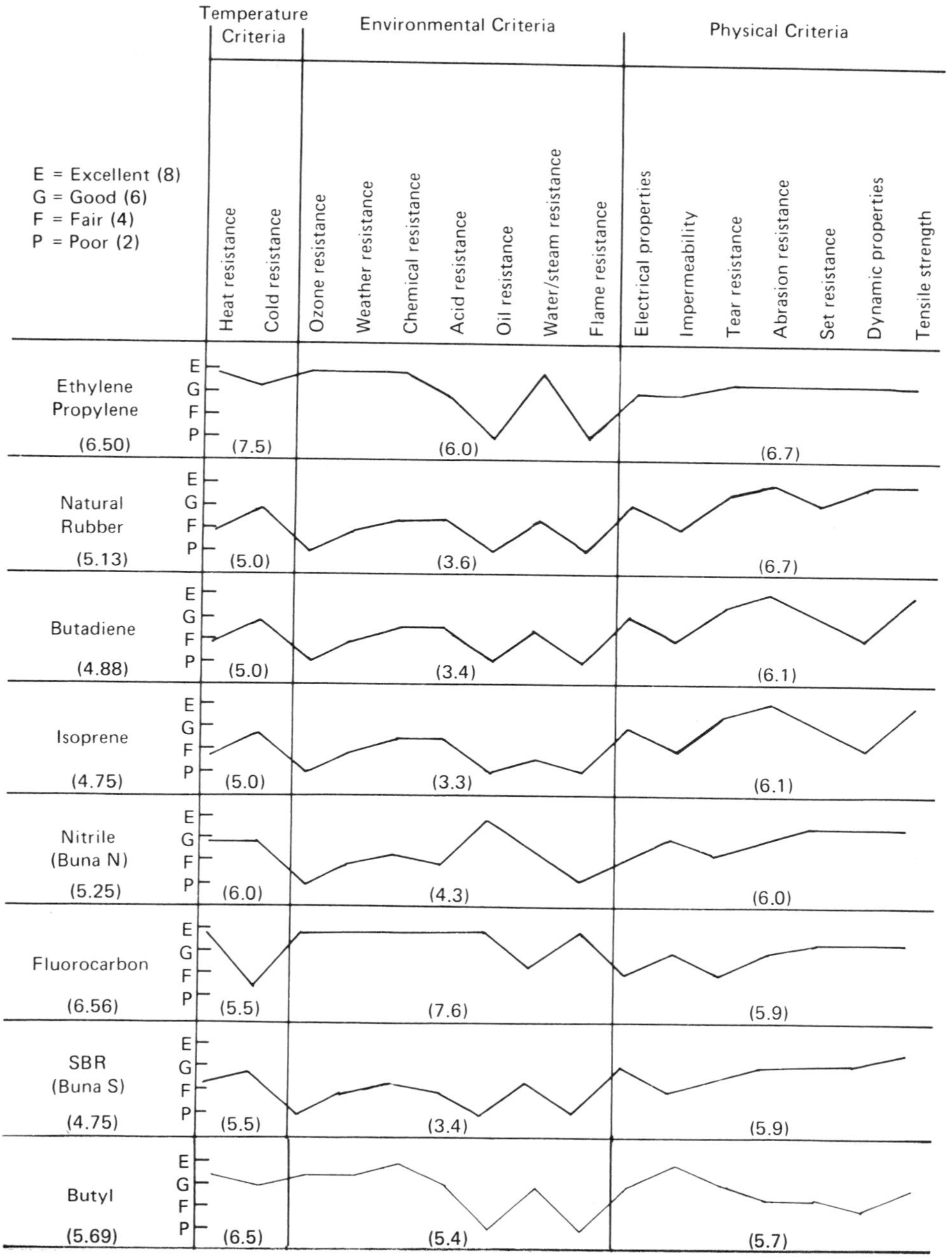

Chart 1D. Elastomers: best physical resistance

61

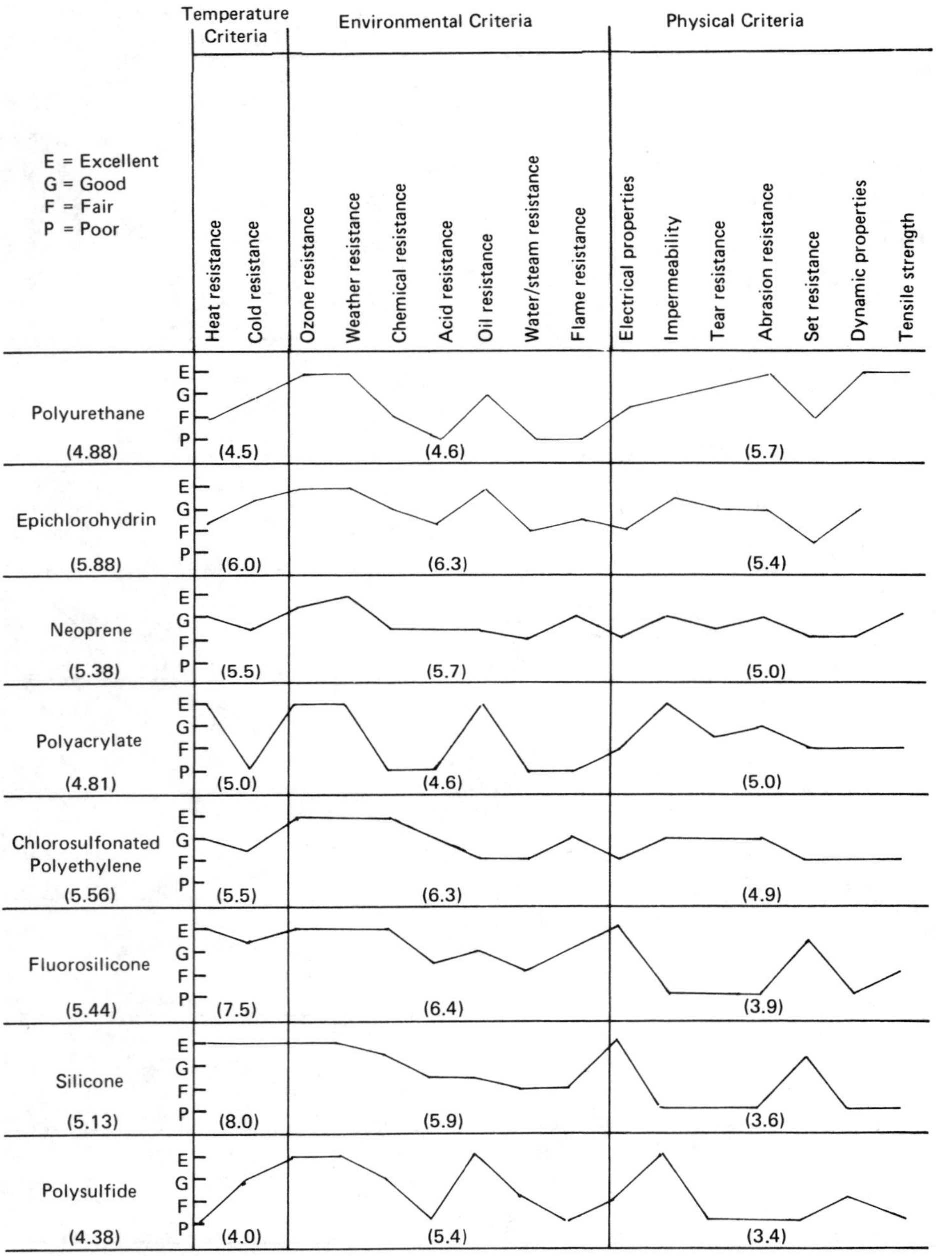

The chart presents rubber material ratings across three criteria groups. The rating scale is:

E = Excellent
G = Good
F = Fair
P = Poor

Material	Temperature Criteria	Environmental Criteria	Physical Criteria
Polyurethane (4.88)	(4.5)	(4.6)	(5.7)
Epichlorohydrin (5.88)	(6.0)	(6.3)	(5.4)
Neoprene (5.38)	(5.5)	(5.7)	(5.0)
Polyacrylate (4.81)	(5.0)	(4.6)	(5.0)
Chlorosulfonated Polyethylene (5.56)	(5.5)	(6.3)	(4.9)
Fluorosilicone (5.44)	(7.5)	(6.4)	(3.9)
Silicone (5.13)	(8.0)	(5.9)	(3.6)
Polysulfide (4.38)	(4.0)	(5.4)	(3.4)

Column headers: Heat resistance, Cold resistance, Ozone resistance, Weather resistance, Chemical resistance, Acid resistance, Oil resistance, Water/steam resistance, Flame resistance, Electrical properties, Impermeability, Tear resistance, Abrasion resistance, Set resistance, Dynamic properties, Tensile strength

Chart 1D (continued)

Table 11. Preferred Material in Descending Order of Overall Performance, Temperature Resistance, Environmental Resistance, and Physical Resistance

Overall performance	Temperature resistance	Environmental resistance	Physical resistance
Fluorocarbon	Silicone	Fluorocarbon	Ethylene propylene
Ethylene propylene	Fluorosilicone	Fluorosilicone	Natural rubber
Epichlorohydrin	Ethylene propylene	Epichlorohydrin	Butadiene
Butyl	Butyl	Chlorosulfonated polyethylene	Isoprene
Chlorosulfonated polyethylene	Epichlorohydrin	Ethylene propylene	Nitrile (Buna N)
Fluorosilicone	Nitrile (Buna N)	Silicone	Fluorocarbon
Neoprene	Fluorocarbon	Neoprene	SBR (Buna S)
Nitrile (Buna N)	Chlorosulfonated polyethylene	Butyl	Butyl
Natural rubber	Neoprene	Polysulfide	Polyurethane
Silicone	SBR (Buna S)	Polyurethane	Epichlorohydrin
Butadiene	Polyacrylate	Polyacrylate	Neoprene
Polyurethane	Nautral rubber	Nitrile (Buna N)	Polyacrylate
Polyacrylate	Butadiene	Nautral rubber	Chlorosulfonated polyethylene
SBR (Buna S)	Isoprene	Butadiene	Fluorosilicone
Isoprene	Polyurethane	SBR (Buna S)	Silicone
Polysulfide	Polysulfide	Isoprene	Polysulfide

have been averaged into an overall weighted equivalent number which appears below the name of the material in the right-hand column of each chart. Here again the designer must use such numbers as an aid in selecting the best material for the particular application. These overall, averaged weighted equivalent numbers have been used to arrange the materials in the right-hand column of Table 11, under overall performance. The materials are also listed in preference under temperature, environmental, and physical resistance. Therefore, silicone is rated as having the best resistance to temperature, while fluorocarbon is most resistant to environmental parameters. Table 11 is essentially a summary of Charts 1A to 1D and may aid the designer in overall performance selection. For example, ethylene propylene is ranked in the upper third of each performance column.

Ethylene propylene has the best overall resistance to physical abuse according to Table 11, but it must be noted that butyl and polysulfide are more resistant to permeability, and fluorosilicone and silicone are more resistant to electrical degradation. Again, the designer should consult all available information; Tables 6, 9, and 11, Charts 1A to 1D, and individual compound and material descriptions presented in Sec. III, Chap. 1.

II
SPECIFIC APPLICATIONS OF RING SEALS

3

Static and Reciprocating Seal Applications: Clearing up the Confusion Between the Military and Industrial O-Ring Gland Specifications

This chapter discusses the confusion between the military and industrial specifications of O-ring gland designs relative to static and reciprocating seal applications. Confusion between the two specifications is attributed predominantly to the fact that the military specification covers both static and reciprocating seal designs without any distinction within the specification. Because reciprocating seals require less O-ring squeeze than static seals in order to function, the military specification inherently specifies less O-ring squeeze than the industrial static O-ring specification.

The material presented in this chapter should be considered when designing O-ring glands for either static or reciprocating seal applications. Additional information on static seal applications is presented in Chap. 4 and additional information on reciprocating seal applications is presented in Chap. 5.

I. GENERAL AREAS OF CONFUSION BETWEEN MILITARY AND INDUSTRIAL SPECIFICATIONS

As the use of O-ring seals have become more prevalent in mechanical systems, a growing confusing has developed in the minds of designers. This confusion centers around the application of military and industrial specifications in the design of O-ring glands. Both military and industrial design specifications incorporate the same standard-size rubber O-ring seals, but in different ways. For example, a size 116 O-ring in a male gland designed in accordance with the industrial specification will experience a maximum stretch of 5.9%, compared to a maximum stretch of only 2.0% when used in a gland designed in accordance with the military specification.

Table 12. Military Gland Design and O-Ring Selection

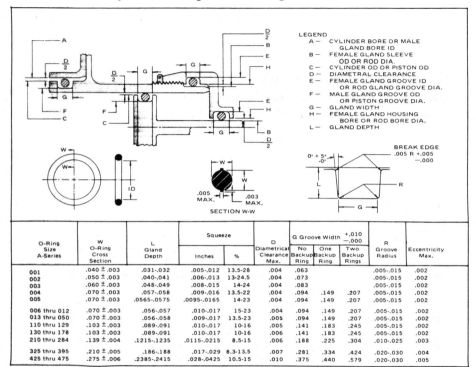

O-Ring Size A-Series	W O-Ring Cross Section	L Gland Depth	Squeeze Inches	Squeeze %	D Diametrical Clearance Max.	G Groove Width +.010 −.000 No Backup Ring	G Groove Width +.010 −.000 One Backup Ring	G Groove Width +.010 −.000 Two Backup Rings	R Groove Radius	Eccentricity Max.
001	.040 ± .003	.031-.032	.005-.012	13.5-28	.004	.063			.005-.015	.002
002	.050 ± .003	.040-.041	.006-.013	13-24.5	.004	.073			.005-.015	.002
003	.060 ± .003	.048-.049	.008-.015	14-24	.004	.083			.005-.015	.002
004	.070 ± .003	.057-.058	.009-.016	13.5-22	.004	.094	.149	.207	.005-.015	.002
005	.070 ± .003	.0565-.0575	.0095-.0165	14-23	.004	.094	.149	.207	.005-.015	.002
006 thru 012	.070 ± .003	.056-.057	.010-.017	15-23	.004	.094	.149	.207	.005-.015	.002
013 thru 050	.070 ± .003	.056-.057	.009-.017	13.5-23	.005	.094	.149	.207	.005-.015	.002
110 thru 129	.103 ± .003	.089-.091	.010-.017	10-16	.005	.141	.183	.245	.005-.015	.002
130 thru 178	.103 ± .003	.089-.091	.010-.017	10-16	.006	.141	.183	.245	.005-.015	.002
210 thru 284	.139 ± .004	.1215-.1235	.0115-.0215	8.5-15	.006	.188	.225	.304	.010-.025	.003
325 thru 395	.210 ± .005	.186-.188	.017-.029	8.3-13.5	.007	.281	.334	.424	.020-.030	.004
425 thru 475	.275 ± .006	.2385-.2415	.028-.0425	10.5-15	.010	.375	.440	.579	.020-.030	.005

Even given the discripancy between the military and industrial specifications, it is up to the designer to determine what specification is applicable to the desired intent of the design. Decisions as to whether a radial gland design is better than a face-seal gland design, when a backup ring should be used, and what is the minimum surface finish required or the proper squeeze can become very confusing when the differences between the two design specifications are not understood. This confusion can be cleared up by analyzing various discrepancies between the military and industrial specifications.

The one major discrepancy that leads to all the confusion between the military and industrial specifications is the fact that the military specification covers dimensions for both static and reciprocating seal designs in the same specification, while the industrial specification is separated into two distinct categories. This re-

Table 12. (continued)

MS28775 SIZE	I.D.	±(1)	W	O.D.	C PISTON DIA	A BORE DIA	F PISTON GROOVE DIA	B ROD DIA	H ROD BORE DIA	E ROD GLAND GROOVE DIA
001	.029	.004	.040	.109	.093	.095	.033	.033	.035	.095
002	.042	.004	.050	.142	.126 +.000	.128 +.001	.048 +.000	.048 +.000	.050 +.001	.128 +.001
003	.056	.004	.060 ±.003	.176	.157 −.001	.159 −.000	.063 −.001	.063 −.001	.065 −.000	.159 −.000
004	.070	.004		.210	.188	.190	.076	.076	.078	.190
005	.101	.004		.241	.219	.221	.108	.108	.110	.221
006	.114	.005		.254	.233	.235	.123	.123	.125	.235
007	.145	.005		.285	.264	.266	.154	.154	.156	.266
008	.176	.005	.070	.316	.295 +.000	.297 +.001	.185 +.000	.185 +.000	.187 +.001	.297 +.001
009	.208	.005	±.003	.348	.327 −.001	.329 −.000	.217 −.001	.217 −.001	.219 −.000	.329 −.000
010	.239	.005		.379	.358	.360	.248	.248	.250	.360
011	.301	.005		.441	.420	.422	.310	.310	.312	.422
012	.364	.005		.504	.483	.485	.373	.373	.375	.485
*013	.426	.005		.566	.548	.550	.438	.435	.437	.547
*014	.489	.005		.629	.611	.613	.501	.498	.500	.610
*015	.551	.005		.691	.673	.675	.563	.560	.562	.672
*016	.614	.005		.754	.736	.738	.626	.623	.625	.735
*017	.676	.005		.816	.798	.800	.688	.685	.687	.797
*018	.739	.005		.879	.861	.863	.751	.748	.750	.860
*019	.801	.006		.941	.923	.925	.813	.810	.812	.922
*020	.864	.006	.070	1.004	.989 +.000	.991 +.002	.879 +.000	.873 +.000	.875 +.001	.985 +.002
*021	.926	.006		1.006	1.051 −.001	1.053 −.000	.941 −.002	.935 −.002	.937 −.000	1.047 −.000
*022	.989	.006	±.003	1.129	1.114	1.116	1.004	.998	1.000	1.110
*023	1.051	.006		1.191	1.176	1.178	1.066	1.060	1.062	1.172
*024	1.114	.006		1.254	1.239	1.241	1.129	1.123	1.125	1.235
*025	1.176	.006		1.316	1.301	1.303	1.191	1.185	1.250	1.297
*026	1.239	.006		1.379	1.364	1.366	1.254	1.248	1.259	1.360
*027	1.301	.006		1.441	1.426	1.428	1.316	1.310	1.312	1.422
*028	1.364	.006		1.504	1.489	1.491	1.379	1.373	1.375	1.485
110	.362	.005		.568	.548	.550	.372	.373	.375	.551
111	.424	.005		.630	.611	.613	.435	.435	.437	.613
112	.487	.005		.693	.673 +.000	.675 +.002	.497 +.000	.498 +.000	.500 +.001	.676 +.002
113	.549	.005	.103	.755	.736 −.001	.738 −.000	.560 −.002	.560 −.002	:562 −.000	.738 −.000
114	.612	.005	±.003	.818	.798	.800	.622	.623	.625	.801
115	.674	.005		.861	.861	.863	.685	.685	.687	.863
116	.737	.005		.943	.923	.925	.747	.748	.750	.926
*117	.799	.006		1.005	.989	.991	.813	.810	.812	.988
*118	.862	.006		1.068	1.051 +.000	1.053 +.002	.875 +.000	.873 +.000	.875 +.001	1.051 +.002
*119	.924	.006	.103	1.130	1.114 −.001	1.116 −.000	.938 −.002	.935 −.002	.937 −.000	1.113 −.000
*120	.987	.006	±.003	1.193	1.176	1.178	1.000	.998	1.000	1.176
*121	1.049	.006		1.255	1.239	1.241	1.063	1.060	1.062	1.238
*122	1.112	.006		1.318	1.301	1.303	1.125	1.123	1.125	1.301

sults in the military specification requiring less (possibly one-half) O-ring squeeze than that specified by the industrial specification for static gland designs. Table 12 presents gland design and O-ring selection according to the military specification, MIL-G-5514F. Notice that it covers male and female radial glands and reciprocating piston and rod glands. Table 13 presents male and female radial gland designs for industrial O-ring static seal

Table 12. (continued)

MS28775 SIZE	I.D.	±(1)	W	O.D.	C PISTON DIA	A BORE DIA	F PISTON GROOVE DIA	B ROD BORE DIA	H ROD DIA	E ROD GLAND GROOVE DIA
*123	1.174	.006		1.380	1.364	1.366	1.188	1.185	1.187	1.363
*124	1.237	.006		1.443	1.426	1.428	1.250	1.248	1.250 +.001	1.426
*125	1.299	.006		1.505	1.489	1.491	1.313	1.310	1.312 −.000	1.488
*126	1.362	.006		1.568	1.551	1.553	1.375	1.373	1.375	1.551
*127	1.424	.006		1.630	1.614	1.616	1.438	1.435	1.437	1.613
*128	1.487	.006		1.693	1.676	1.678	1.500	1.498	1.500	1.676
*129	1.549	.010		1.755	1.739	1.741	1.563	1.560	1.562	1.738
*130	1.612	.010		1.818	1.802	1.805	1.627	1.623	1.625	1.801
*131	1.674	.010		1.880	1.864	1.867	1.689	1.685	1.687	1.863
*132	1.737	.010		1.943	1.927	1.930	1.752	1.748	1.750	1.926
*133	1.799	.010		2.005	1.989	1.992	1.814	1.810	1.813	1.988
*134	1.862	.010		2.068	2.052	2.055	1.877	1.873	1.876	2.051
*135	1.925	.010	.103	2.131	2.115 +.000	2.118 +.002	1.940 +.000	1.936 +.000	1.939 +.002	2.114 +.002
*136	1.987	.010	±.003	2.193	2.117 −.001	2.180 −.000	2.002 −.002	1.998 −.002	2.001 −.000	2.176 −.000
*137	2.050	.010		2.256	2.240	2.243	2.065	2.061	2.064	2.239
*138	2.112	.010		2.318	2.302	2.305	2.127	2.123	2.126	2.301
*139	2.175	.010		2.381	2.365	2.368	2.190	2.186	2.189	2.364
*140	2.237	.010		2.443	2.427	2.430	2.252	2.248	2.251	2.426
*141	2.300	.010		2.506	2.490	2.493	2.315	2.311	2.314	2.489
*142	2.362	.010		2.568	2.552	2.555	2.377	2.373	2.376	2.551
*143	2.425	.010		2.631	2.615 +.000	2.618	2.440	2.436	2.439	2.614
*144	2.487	.010		2.693	2.677 −.002	2.680	2.502	2.498	2.501	2.676
*145	2.550	.010		2.756	2.740	2.743	2.565	2.561	2.564	2.739
*146	2.612	.010		2.818	2.802	2.805	2.627	2.623	2.626	2.801
*147	2.675	.015		2.881	2.865	2.868	2.690	2.686	2.689	2.864
*148	2.737	.015		2.943	2.927	2.930	2.752	2.748	2.751	2.926
*149	2.800	.015		3.006	2.990	2.993	2.815	2.811	2.814	2.989
210	.734	.006		1.012	.989	.991	.748	.748	.750	.991
211	.796	.006		1.074	1.051	1.053	.810	.810	.812	1.053
212	.859	.006		1.137	1.114	1.116	.873	.873	.875	1.116
213	.921	.006	.139	1.199	1.176 +.000	1.178 +.002	.935 +.000	.935 +.000	.937 +.001	1.178 +.002
214	.984	.006	±.004	1.262	1.239 −.001	1.241 −.000	.998 −.002	.998 −.002	1.000 −.000	1.241 −.000
215	1.046	.006		1.324	1.301	1.303	1.063	1.060	1.062	1.303
216	1.109	.006		1.387	1.364	1.366	1.123	1.123	1.125	1.366
217	1.171	.006		1.449	1.426	1.428	1.185	1.185	1.187	1.428
218	1.234	.006		1.512	1.489	1.491	1.248	1.248	1.250	1.491
219	1.296	.006		1.574	1.551	1.553	1.310	1.310	1.312	1.553
220	1.359	.006		1.637	1.614	1.616	1.373	1.373	1.375	1.616
221	1.421	.006		1.699	1.676	1.678	1.435	1.435	1.437	1.678
222	1.484	.006		1.762	1.739	1.741	1.498	1.498	1.500	1.741
*223	1.609	.010	.139	1.887	1.864 +.000	1.867 +.002	1.624 +.000	1.623 +.000	1.625 +.002	1.866 +.002
*224	1.734	.010	±.004	2.012	1.989 −.001	1.992 −.000	1.749 −.002	1.748 −.002	1.750 −.000	1.991 −.000

Table 12. (continued)

MS28775 SIZE	I.D.	±(1)	W	O.D.	C PISTON DIA	A BORE DIA	F PISTON GROOVE DIA	B ROD BORE DIA	H ROD DIA	E ROD GLAND GROOVE DIA
*225	1.859	.010		2.137	2.115 +.000	2.118	1.875	1.873	1.876	2.116
*226	1.984	.010		2.262	2.240 −.001	2.243	2.000	1.998	2.001	2.241
*227	2.109	.010		2.387	2.365	2.368	2.125	2.123	2.126	2.366
*228	2.234	.010		2.512	2.490	2.493	2.250	2.248	2.251	2.491
*229	2.359	.010		2.637	2.615	2.618	2.375	2.373	2.376	2.616
*230	2.484	.010		2.763	2.740	2.743	2.500	2.498	2.501	2.741
*231	2.609	.010		2.887	2.865	2.868	2.625	2.623	2.626	2.866
*232	2.734	.015		3.012	2.990	2.993	2.750	2.748	2.751	2.991
*233	2.859	.015		3.137	3.115	3.118	2.875	2.873	2.876	3.116
*234	2.984	.015		3.262	3.240	3.243	3.000	2.997	3.000	3.240
*235	3.109	.015		3.387	3.365	3.368	3.125	3.122	3.125	3.365
*236	3.234	.015	.139	3.512	3.490 +.000	3.493 +.002	3.250 +.000	3.247 +.000	3.250 +.002	3.490 +.002
*237	3.359	.015	±.004	3.637	3.615 −.002	3.618 −.000	3.375 −.002	3.372 −.002	3.375 −.000	3.615 −.000
*238	3.484	.015		3.762	3.740	3.743	3.500	3.497	3.500	3.740
*239	3.609	.015		3.887	3.865	3.868	3.625	3.622	3.625	3.865
*240	3.734	.015		4.012	3.990	3.993	3.750	3.747	3.750	3.990
*241	3.859	.015		4.137	4.115	4.118	3.875	3.872	3.875	4.115
*242	3.984	.015		4.262	4.240	4.243	4.000	3.997	4.000	4.240
*243	4.109	.015		4.387	4.365	4.368	4.125	4.122	4.125	4.365
*244	4.234	.015		4.512	4.489	4.493	4.250	4.247	4.250	4.490
*245	4.359	.015		4.637	4.614	4.618	4.375	4.372	4.375	4.615
*246	4.484	.015		4.762	4.739	4.743	4.500	4.497	4.501	4.740
*247	4.609	.015		4.887	4.864	4.868	4.625	4.622	4.626	4.865
325	1.475	.010		1.895	1.864	1.867	1.495	1.498	1.500	1.870
326	1.600	.010		2.020	1.989 +.000	1.992	1.620	1.623	1.625	1.995
327	1.725	.010		2.145	2.115 −.001	2.118	1.746	1.748	1.750	2.120
328	1.850	.010		2.270	2.240	2.243	1.871	1.873	1.876	2.245
329	1.975	.010		2.395	2.365	2.368	2.996	1.998	2.001	2.370
330	2.100	.010		2.520	2.490	2.493	2.121	2.123	2.126	2.495
331	2.225	.010		2.645	2.615	2.618	2.246	2.248	2.251	2.620
332	2.350	.010		2.770	2.740	2.743	2.371	2.373	2.376	2.745
333	2.475	.010	.210	2.895	2.865 +.000	2.868 +.002	2.496 +.000	2.498 +.000	2.501 +.002	2.870 +.002
334	2.600	.010	±.005	3.020	2.990 −.002	2.993 +.000	2.621 −.002	2.623 −.002	2.626 −.000	2.995 −.000
335	2.725	.015		3.145	3.115	3.118	2.746	2.748	2.751	3.120
336	2.850	.015		3.270	3.240	3.243	2.871	2.873	2.876	3.245
337	2.975	.015		3.395	3.365	3.368	2.996	2.997	3.000	3.369
338	3.100	.015		3.520	3.490	3.493	3.121	3.122	3.125	3.494
339	3.225	.015		3.645	3.615	3.618	3.246	3.247	3.250	3.619
340	3.350	.015		3.770	3.740	3.743	3.371	3.372	3.375	3.744
341	3.475	.015		3.895	3.865	3.868	3.496	3.497	3.500	3.869
342	3.600	.015		4.020	3.990	3.993	3.621	3.622	3.625	3.994
343	3.725	.015		4.145	4.115	4.118	3.746	3.747	3.750	4.119

Table 12. (continued)

MS28775 SIZE A	DIMENSIONS				C PISTON DIA		A BORE DIA		F PISTON GROOVE DIA		B ROD BORE DIA		H ROD DIA		E ROD GLAND GROOVE DIA	
	I.D.	±(1)	W	O.D.												
344	3.850	.015		4.270	4.240		4.243		3.871		3.872		3.875		4.244	
345	3.975	.015		4.395	4.365		4.368		3.996		3.997		4.000		4.369	
346	4.100	.015	.210	4.520	4.489	+.000	4.493	+.002	4.121	+.000	4.122	+.000	4.125	+.002	4.494	+.002
347	4.225	.015	±.005	4.645	4.614	−.002	4.618	−.000	4.246	−.002	4.247	−.002	4.250	−.000	4.619	−.000
348	4.350	.015		4.770	4.739		4.743		4.371		4.372		4.375		4.744	
349	4.475	.015		4.895	4.864		4.868		4.496		4.497		4.500		4.869	
425	4.475	.015		5.025	4.970		4.974		4.497		4.497		4.501		4.974	
426	4.600	.015		5.150	5.095		5.099		4.622		4.622		4.626		5.099	
427	4.725	.015		5.275	5.220		5.224		4.747		4.747		4.751		5.224	
428	4.850	.015		5.400	5.345		5.349		4.872		4.872		4.876		5.349	
429	4.975	.015		5.525	5.470		5.474		4.997		4.997		5.001		5.474	
430	5.100	.023		5.650	5.595		5.599		5.122		5.122		5.126		5.599	
431	5.225	.023		5.775	5.720		5.724		5.247		5.247		5.251		5.724	
432	5.350	.023		5.900	5.845		5.849		5.372		5.372		5.376		5.849	
433	5.475	.023		6.025	5.970		5.974		5.497		5.497		5.501		5.974	
434	5.600	.023		6.150	6.095		6.099		5.622		5.622		5.626		6.099	
435	5.725	.023		6.275	6.220		6.224		5.747		5.747		5.751	+.002	6.224	
436	5.850	.023		6.400	6.345		6.349		5.872		5.872		5.867	−.000	6.349	
437	5.975	.023		6.525	6.470		6.474		5.997		5.997		5.501		6.474	
438	6.225	.023	.275	6.775	6.720	+.000	6.724	+.003	6.247	+.000	6.247	+.000	6.251		6.724	+.003
439	6.475	.023	±.006	7.025	6.970	−.002	6.974	−.000	6.497	−.003	6.497	−.003	6.501		6.974	−.000
440	6.725	.023		7.275	7.220		7.224		6.747		6.747		6.751		7.224	
441	6.975	.023		7.525	7.470		7.474		6.997		6.997		7.001		7.474	
442	7.225	.030		7.775	7.720		7.724		7.247		7.247		7.251		7.724	
443	7.475	.030		8.025	7.970		7.974		7.497		7.497		7.501		7.974	
444	7.725	.030		8.275	8.220		8.224		7.747		7.747		7.751		8.244	
445	7.975	.030		8.525	8.470		8.474		7.997		7.997		8.001		8.474	
446	8.475	.030		9.025	8.970		8.974		8.497		8.498		8.501		8.974	
447	8.975	.030		9.525	9.470		9.474		8.997		8.997		9.001		9.474	
448	9.475	.030		10.025	9.970		9.974		9.497		9.497		9.501	+.003	9.974	
449	9.975	.030		10.525	10.470		10.474		9.997		9.997		10.001	−.000	10.474	
450	10.475	.030		11.025	10.970		10.974		10.497		10.497		10.501		10.974	
451	10.975	.030		11.525	11.470		11.474		10.997		10.997		11.001		11.474	
452	11.475	.030		12.025	11.970	+.000	11.974	+.004	11.497		11.497		11.501		11.974	
453	11.975	.030		12.525	12.470	−.003	12.474	−.000	11.997		11.997		12.001		12.474	−.004
454	12.475	.030		13.025	12.970		12.974		12.497		12.497		12.501		12.974	−.000
455	12.975	.030		13.525	13.470		13.474		12.997		12.997		13.001		13.474	
456	13.475	.030		14.025	13.970		13.974		13.497		13.497		13.501		13.974	
457	13.975	.030		14.525	14.470		14.474		13.997		13.997		14.001		14.474	
458	14.475	.030		15.025	14.970		14.974		14.497		14.497		14.501		14.974	
459	14.975	.030		15.525	15.470		15.474		14.997		14.997		15.001		15.474	
460	15.475	.030		16.025	15.970		15.974		15.497		15.497		15.501		15.974	

*Static applications only.

(1) Inside diameter tolerances for O-rings with inside diameters over ½ in. have been increased per AS568A.

Source: O-Ring Design and Selection Handbook 110-A, Sargent Industries, Carlsbad, Calif., 1976.

Table 13. Industrial O-Ring Static Seal Glands

A. Gland Details and Design Chart

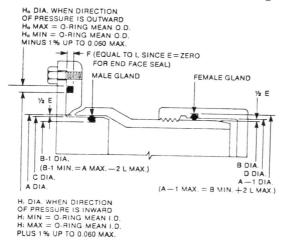

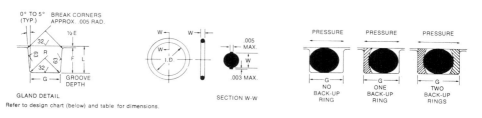

GLAND DETAIL
Refer to design chart (below) and table for dimensions.

SECTION W-W

O-RING SIZE PARKER 2-	W CROSS SECTION		L GLAND DEPTH	SQUEEZE		E (a) (c) DIAMETRAL CLEARANCE	G GROOVE WIDTH			R GROOVE RADIUS	ECCENTRICITY MAX. (b)
	Nominal	Actual		ACTUAL	%		NO BACK-UP RINGS	ONE BACK-UP RING	TWO BACK-UP RINGS		
004 through 050	1/16	.070 ±.003	.050 to .052	.015 to .023	22 to 32	.002 to .005	.093 to .098	.138 to .143	.205 to .210	.005 to .015	.002
102 through 178	3/32	.103 ±.003	.081 to .083	.017 to .025	17 to 24	.002 to .005	.140 to .145	.171 to .176	.238 to .243	.005 to .015	.002
201 through 284	1/8	.139 ±.004	.111 to .113	.022 to .032	16 to 23	.003 to .006	.187 to .192	208 to .213	.275 to .280	.010 to .025	.003
309 through 395	3/16	.210 ±.005	.170 to .173	.032 to .045	15 to 21	.003 to .006	.281 to .286	.311 to .316	.410 to .415	.020 to .035	.004
425 through 475	1/4	.275 ±.006	.226 to .229	.040 to .055	15 to 20	.004 to .007	.375 to .380	.408 to .413	.538 to .543	.020 to .035	.005

[a]Clearance gap must be held to a minimum consistent with design requirements for temperature range variation.
[b]Total Indicator reading between groove and adjacent bearing surface.
[c]Reduce maximum diametral clearance 50% when using silicone O- rings.
Source: Reproduced by permission of Parker Seal Company.

Table 13. (continued)

B. Gland Dimensions (1500 psi Max.)

Parker No. 2-	ID	±(1)	W	Mean OD (Ref)	A BORE DIA. (male gland) +.002/-.000	A-1 GROOVE DIA. (female gland) -.000/+	B TUBE OD (female gland) +.000/-.002	B-1 GROOVE DIA. (male gland) +.000/-	C PLUG DIA. (male gland) +.000/-.001	D THROAT DIA. (female gland) +.001/-.000	G† GROOVE WIDTH	H OD FACE GROOVE (Use for Internal Pressure only) +.000/-		F DEPTH FACE GROOVE
5-051	.070	.005	.040	.150	.146	.142	.081	.085	*.144	.083	.055	.150	.002	.0305 +.002
2-001	.029	.004	.040	.109	.105	.101	.040	.044	*.103	.042	.055	.109	.002	.0305 +.002
002	.042	.004	.050	.142	.138	.132 (.002)	.053	.059 (.002)	*.136	.055	.070	.142	.002	.0395 +.002
003	.056	.004	.060	.176	.172	.162	.067	.077	*.170	.069	.083	.176	.002	.0475 +.002
004	.070	.004		.210	.206	.181	.081	.106	*.204	.083		.210	.002	
005	.101	.004		.241	.237	.212	.112	.137	*.235	.114		.241	.002	
006	.114	.005		.254	.250	.225	.125	.150	*.248	.127		.254	.003	
007	.145	.005		.285	.281	.256	.156	.181	*.279	.158		.285	.003	
008	.176	.005		.316	.312	.287	.187	.212	*.310	.189		.316	.003	
009	.208	.005		.348	.343	.318	.218	.243	*.341	.220		.348	.003	
010	.239	.005		.379	.375	.350	.250	.275	*.373	.252		.379	.004	
011	.301	.005		.441	.437	.412	.312	.337	*.435	.314		.441	.004	
012	.364	.005		.504	.500	.475	.375	.400	*.498	.377		.504	.005	
013	.426	.005		.566	.562	.537	.437	.462	.560	.439		.566	.006	
014	.489	.005		.629	.625	.600	.500	.525	.623	.502		.629	.006	
015	.551	.005		.691	.687	.662	.562	.587	.685	.564		.691	.007	
016	.614	.005		.754	.750	.725	.625	.650	.748	.627		.754	.008	
017	.676	.005		.816	.812	.787	.687	.712	.810	.689		.816	.008	
018	.739	.005		.879	.875	.850	.750	.775	.873	.752		.879	.009	
019	.801	.005		.941	.937	.912	.812	.837	.935	.814		.941	.009	
020	.864	.006		1.004	1.000	.975	.875	.900	.998	.877		1.004	.010	
021	.926	.006		1.066	1.062	1.037	.937	.962	1.060	.939	.093	1.066	.011	.050
022	.989	.006	.070	1.129	1.125	1.100 (.002)	1.000	1.025 (.002)	1.123	1.002	+.005	1.129	.011	+.002
023	1.051	.006 ±.003		1.191	1.187	1.162	1.062	1.087	1.185	1.064	-.000	1.191	.012	-.000
024	1.114	.006		1.254	1.250	1.225	1.125	1.150	1.248	1.127		1.254	.013	
025	1.176	.006		1.316	1.312	1.287	1.187	1.212	1.310	1.189		1.316	.013	
026	1.239	.006		1.379	1.375	1.350	1.250	1.275	1.373	1.252		1.379	.014	
027	1.301	.006		1.441	1.437	1.412	1.312	1.337	1.435	1.314		1.441	.014	
028	1.364	.006		1.504	1.500	1.475	1.375	1.400	1.498	1.377		1.504	.015	
029	1.489	.010		1.629	1.625	1.600	1.500	1.525	1.623	1.502		1.629	.016	
030	1.614	.010		1.754	1.750	1.725	1.625	1.650	1.748	1.627		1.754	.018	
031	1.739	.010		1.879	1.875	1.850	1.750	1.775	1.873	1.752		1.879	.019	
032	1.864	.010		2.004	2.000	1.975	1.875	1.900	1.998	1.877		2.004	.020	
033	1.989	.010		2.129	2.125	2.100	2.000	2.025	2.123	2.002		2.129	.021	
034	2.114	.010		2.254	2.250	2.225	2.125	2.150	2.248	2.127		2.254	.023	
035	2.239	.010		2.379	2.375	2.350	2.250	2.275	2.373	2.252		2.379	.024	
036	2.364	.010		2.504	2.500	2.475	2.375	2.400	2.498	2.377		2.504	.025	
037	2.489	.010		2.629	2.625	2.600	2.500	2.525	2.623	2.502		2.629	.026	
038	2.614	.010		2.754	2.750	2.725	2.625	2.650	2.748	2.627		2.754	.028	
039	2.739	.015		2.879	2.875	2.850	2.750	2.775	2.873	2.752		2.879	.029	
040	2.864	.015		3.004	3.000	2.975	2.875	2.900	2.998	2.877		3.004	.030	
041	2.989	.015		3.129	3.125	3.100	3.000	3.025	3.123	3.002		3.129	.031	
042	3.239	.015		3.379	3.375	3.350	3.250	3.275	3.373	3.252		3.379	.034	
043	3.489	.015		3.629	3.625	3.600	3.500	3.525	3.623	3.502		3.629	.036	

Table 13. (continued)

Parker No. 2-	ID	±(1)	W	Mean OD (Ref)	A BORE DIA. (male gland) +.002/-.000	A-1 GROOVE DIA. (female gland) -.000	+	B TUBE OD (female gland) +.000/-.002	B-1 GROOVE DIA. (male gland) +.000	-	C PLUG DIA. (male gland) +.000/-.001	D THROAT DIA. (female gland) +.001/-.000	G† GROOVE WIDTH	H₀ OD FACE GROOVE +.000		F DEPTH FACE GROOVE -
044	3.739	.015	.070	3.879	3.875	3.850		3.750	3.775		3.873	3.752		3.879	.039	
045	3.989	.015		4.129	4.125	4.100	.002	4.000	4.025	.002	4.123	4.002	.093	4.129	.041	.050
046	4.239	.015	±.003	4.379	4.375	4.350		4.250	4.275		4.373	4.252	+.005	4.379	.044	+.002
047	4.489	.015		4.629	4.625	4.600		4.500	4.525		4.623	4.502	-.000	4.629	.046	-.000
048	4.739	.015		4.879	4.875	4.850		4.750	4.775		4.873	4.752		4.879	.049	
049	4.989	.023		5.129	5.125	5.100		5.000	5.025		5.123	5.002		5.129	.051	
050	5.239	.023		5.379	5.375	5.350		5.250	5.275		5.373	5.252		5.379	.054	
102	.049	.004		.255	.247	.224		.062	.085		* .245	.064		.255	.003	
103	.081	.005		.287	.278	.256		.094	.116		* .276	.095		.287	.003	
104	.112	.005		.318	.310	.287		.125	.148		* .308	.127		.318	.003	
105	.143	.005		.349	.342	.318		.156	.180		* .340	.158		.349	.003	
106	.174	.005		.380	.374	.349		.187	.212		* .372	.189		.380	.004	
107	.206	.005		.412	.405	.381		.219	.243		* .403	.221		.412	.004	
108	.237	.005		.443	.437	.412		.250	.275		* .435	.252		.443	.004	
109	.299	.005		.505	.500	.474		.312	.338		* .498	.314		.505	.005	
110	.362	.005		.568	.562	.537		.375	.400		* .560	.377		.568	.006	
111	.424	.005		.630	.625	.599		.437	.463		* .623	.439		.630	.006	
112	.487	.005		.693	.687	.662		.500	.525		* .685	.502		.693	.007	
113	.549	.005		.755	.750	.724		.562	.588		* .748	.564		.755	.008	
114	.612	.005		.818	.812	.787		.625	.650		.810	.627		.818	.008	
115	.674	.005		.880	.875	.849		.687	.713		.873	.689		.880	.009	
116	.737	.005		.943	.937	.912		.750	.775		.935	.752		.943	.009	
117	.799	.006		1.005	1.000	.974		.812	.838		.998	.814		1.005	.010	
118	.862	.006	.103	1.068	1.062	1.037		.875	.900		1.060	.877	.140	1.068	.011	.081
119	.924	.006		1.130	1.125	1.099	.002	.937	.963	.002	1.123	.939	+.005	1.130	.011	+.002
120	.987	.006	±.003	1.193	1.187	1.162		1.000	1.025		1.185	1.002	-.000	1.193	.012	-.000
121	1.049	.006		1.255	1.250	1.224		1.062	1.088		1.248	1.064		1.255	.013	
122	1.112	.006		1.318	1.312	1.287		1.125	1.150		1.310	1.127		1.318	.013	
123	1.174	.006		1.380	1.375	1.349		1.187	1.213		1.373	1.189		1.380	.014	
124	1.237	.006		1.443	1.437	1.412		1.250	1.275		1.435	1.252		1.443	.014	
125	1.299	.006		1.505	1.500	1.474		1.312	1.338		1.498	1.314		1.505	.015	
126	1.362	.006		1.568	1.562	1.537		1.375	1.400		1.560	1.377		1.568	.016	
127	1.424	.006		1.630	1.625	1.599		1.437	1.463		1.623	1.439		1.630	.016	
128	1.487	.010		1.693	1.687	1.662		1.500	1.525		1.685	1.502		1.693	.017	
129	1.549	.010		1.755	1.750	1.724		1.562	1.588		1.748	1.564		1.755	.017	
130	1.612	.010		1.818	1.812	1.787		1.625	1.650		1.810	1.627		1.818	.018	
131	1.674	.010		1.880	1.875	1.849		1.687	1.713		1.873	1.689		1.880	.019	
132	1.737	.010		1.943	1.937	1.912		1.750	1.775		1.935	1.752		1.943	.019	
133	1.799	.010		2.005	2.000	1.974		1.812	1.838		1.998	1.814		2.005	.020	
134	1.862	.010		2.068	2.062	2.037		1.875	1.900		2.060	1.877		2.068	.021	
135	1.925	.010		2.131	2.125	2.099		1.937	1.963		2.123	1.939		2.131	.021	
136	1.987	.010		2.193	2.187	2.162		2.000	2.025		2.185	2.002		2.193	.022	
137	2.050	.010		2.256	2.250	2.224		2.062	2.088		2.248	2.064		2.256	.023	

Table 13. (continued)

O-Ring Size Parker No. 2-	ID	±(1)	W	Mean OD (Ref) +.002 -.000	A BORE DIA. (male gland)	A-1 GROOVE DIA. (female gland) -.000	+	B TUBE OD (female gland) +.000 -.002	B-1 GROOVE DIA. (male gland) +.000	-	C PLUG DIA. (male gland) +.000 -.001	D THROAT DIA. (female gland) +.001 -.000	Gt GROOVE WIDTH	Ho OD FACE GROOVE (Use for Internal Pressure only) +.000	F DEPTH FACE GROOVE -
138	2.112	.010		2.318	2.312	2.287		2.125	2.150		2.310	2.127		2.318 .023	
139	2.175	.010		2.381	2.375	2.349		2.187	2.213		2.373	2.189		2.381 .024	
140	2.237	.010		2.443	2.437	2.412		2.250	2.275		2.435	2.252		2.443 .024	
141	2.300	.010		2.506	2.500	2.474		2.312	2.338		2.498	2.315		2.506 .025	
142	2.362	.010		2.568	2.562	2.537		2.375	2.400		2.560	2.377		2.568 .026	
143	2.425	.010		2.631	2.625	2.599		2.437	2.463		2.623	2.439		2.631 .026	
144	2.487	.010		2.693	2.687	2.662		2.500	2.525		2.685	2.502		2.693 .027	
145	2.550	.010		2.756	2.750	2.724		2.562	2.588		2.748	2.564		2.756 .028	
146	2.612	.010		2.818	2.812	2.787		2.625	2.650		2.810	2.627		2.818 .028	
147	2.675	.015		2.881	2.875	2.849		2.687	2.713		2.873	2.689		2.881 .029	
148	2.737	.015		2.943	2.937	2.912		2.750	2.775		2.935	2.752		2.943 .029	
149	2.800	.015		3.006	3.000	2.974		2.812	2.838		2.998	2.814		3.006 .030	
150	2.862	.015		3.068	3.062	3.037		2.875	2.900		3.060	2.877		3.068 .031	
151	2.987	.015		3.193	3.187	3.162		3.000	3.025		3.185	3.002		3.193 .032	
152	3.237	.015		3.443	3.437	3.412		3.250	3.275		3.435	3.252		3.443 .034	
153	3.487	.015		3.693	3.687	3.662	.002	3.500	3.525	.002	3.685	3.502		3.693 .037	
154	3.737	.015	.103	3.943	3.937	3.912		3.750	3.775		3.935	3.752	.140	3.943 .039	.081
155	3.987	.015	±.003	4.193	4.187	4.162		4.000	4.025		4.185	4.002	+.005	4.193 .042	+.002
156	4.237	.015		4.443	4.437	4.412		4.250	4.275		4.435	4.252	-.000	4.443 .044	-.000
157	4.487	.015		4.693	4.687	4.662		4.500	4.525		4.685	4.502		4.693 .047	
158	4.737	.015		4.943	4.937	4.912		4.750	4.775		4.935	4.752		4.943 .049	
159	4.987	.015		5.193	5.187	5.162		5.000	5.025		5.185	5.002		5.193 .052	
160	5.237	.023		5.443	5.437	5.412		5.250	5.275		5.435	5.252		5.443 .054	
161	5.487	.023		5.693	5.687	5.662		5.500	5.525		5.685	5.502		5.693 .057	
162	5.737	.023		5.943	5.937	5.912		5.750	5.775		5.935	5.752		5.943 .060	
163	5.987	.023		6.193	6.187	6.162		6.000	6.025		6.185	6.002		6.193 .060	
164	6.237	.023		6.443	6.437	6.412		6.250	6.275		6.435	6.252		6.443 .060	
165	6.487	.023		6.693	6.687	6.662		6.500	6.525		6.685	6.502		6.693 .060	
166	6.737	.023		6.943	6.937	6.912		6.750	6.775		6.935	6.752		6.943 .060	
167	6.987	.023		7.193	7.187	7.162		7.000	7.025		7.185	7.002		7.193 .060	
168	7.237	.030		7.443	7.437	7.412		7.250	7.275		7.435	7.252		7.443 .060	
169	7.487	.030		7.693	7.687	7.662		7.500	7.525		7.685	7.502		7.693 .060	
170	7.737	.030		7.943	7.937	7.912		7.750	7.775		7.935	7.752		7.943 .060	
171	7.987	.030		8.193	8.187	8.162		8.000	8.025		8.185	8.002		8.193 .060	
172	8.237	.030		8.443	8.437	8.412		8.250	8.275		8.435	8.252		8.443 .060	
173	8.487	.030		8.693	8.687	8.662		8.500	8.525		8.685	8.502		8.693 .060	
174	8.737	.030		8.943	8.937	8.912		8.750	8.775		8.935	8.752		8.943 .060	
175	8.987	.030		9.193	9.187	9.162		9.000	9.025		9.185	9.002		9.193 .060	
176	9.237	.030		9.443	9.437	9.412		9.250	9.275		9.435	9.252		9.443 .060	
177	9.487	.030		9.693	9.687	9.662		9.500	9.525		9.685	9.502		9.693 .060	
178	9.737	.030		9.943	9.937	9.912		9.750	9.775		9.935	9.752		9.943 .060	
201	.171	.005	.139	.449	.437	.409		.187	.215		* .434	.190	.187	.449 .004	.111
202	.234	.005	±.004	.512	.500	.472	.002	.250	.278	.002	* .497	.253	+.005	.512 .005	+.002
203	.296	.005		.574	.562	.534		.312	.340		* .559	.315	-.000	.574 .006	-.000

Table 13. (continued)

Parker No. 2–	ID	±(1)	W	Mean OD (Ref)	A BORE DIA. (male gland) +.002/-.000	A-1 GROOVE DIA. (female gland) -.000	+	B TUBE OD (female gland) +.000/-.002	B-1 GROOVE DIA. (male gland) +.000	-	C PLUG DIA. (male gland) +.000/-.001	D THROAT DIA. (female gland) +.001/-.000	Qt GROOVE WIDTH	H₀ OD FACE GROOVE (Use for Internal Pressure only) +.000	-	F DEPTH FACE GROOVE
204	.359	.005		.637	.625	.597		.375	.403		.622	.378		.637		.006
205	.421	.005		.699	.687	.659		.437	.465		.684	.440		.699		.007
206	.484	.005		.762	.750	.722		.500	.528		.747	.503		.762		.008
207	.546	.005		.824	.812	.784		.562	.590		.809	.565		.824		.008
208	.609	.005		.887	.875	.847		.625	.653		.872	.628		.887		.009
209	.671	.005		.949	.937	.909		.687	.715		.934	.690		.949		.009
210	.734	.006		1.012	1.000	.972		.750	.778		.997	.753		1.012		0.10
211	.796	.006		1.074	1.062	1.034		.812	.840		1.059	.815		1.074		.011
212	.859	.006		1.137	1.125	1.097		.875	.903		1.122	.878		1.137		.011
213	.921	.006		1.199	1.187	1.159		.937	.965		1.184	.940		1.199		.012
214	.984	.006		1.262	1.250	1.222		1.000	1.028		1.247	1.003		1.262		.013
215	1.046	.006		1.324	1.312	1.284		1.062	1.090		1.309	1.065		1.324		.013
216	1.109	.006		1.387	1.375	1.347		1.125	1.153		1.372	1.128		1.387		.014
217	1.171	.006		1.449	1.437	1.409		1.187	1.215		1.434	1.190		1.499		.014
218	1.234	.006		1.512	1.500	1.472		1.250	1.278		1.497	1.253		1.512		.015
219	1.296	.006	.139	1.574	1.562	1.534	.002	1.312	1.340	.002	1.559	1.315	.187	1.574		.015 (.111)
220	1.359	.006		1.637	1.625	1.597		1.375	1.403		1.622	1.378		1.637		.016
221	1.421	.006	±.004	1.700	1.687	1.659		1.437	1.465		1.684	1.440	+.005	1.700		.017 (+.002)
222	1.484	.006		1.762	1.750	1.722		1.500	1.528		1.747	1.503	-.000	1.762		.018 (-.000)
223	1.609	.010		1.887	1.875	1.847		1.625	1.653		1.872	1.628		1.887		.019
224	1.734	.010		2.012	2.000	1.972		1.750	1.778		1.997	1.753		2.012		.020
225	1.859	.010		2.137	2.125	2.097		1.875	1.903		2.122	1.878		2.137		.021
226	1.984	.010		2.262	2.250	2.222		2.000	2.028		2.247	2.003		2.262		.023
227	2.109	.010		2.387	2.375	2.347		2.125	2.153		2.372	2.128		2.387		.024
228	2.234	.010		2.512	2.500	2.472		2.250	2.278		2.497	2.253		2.512		.025
229	2.359	.010		2.637	2.625	2.597		2.375	2.403		2.622	2.378		2.637		.026
230	2.484	.010		2.762	2.750	2.722		2.500	2.528		2.747	2.503		2.762		.028
231	2.609	.010		2.887	2.875	2.847		2.625	2.653		2.872	2.628		2.887		.028
232	2.734	.015		3.012	3.000	2.972		2.750	2.778		2.997	2.753		3.012		.030
233	2.859	.015		3.137	3.125	3.097		2.875	2.903		3.122	2.878		3.137		.031
234	2.984	.015		3.262	3.250	3.222		3.000	3.028		3.247	3.003		3.262		.033
235	3.109	.015		3.387	3.375	3.347		3.125	3.153		3.372	3.128		3.387		.034
236	3.234	.015		3.512	3.500	3.472		3.250	3.278		3.497	3.253		3.512		.035
237	3.359	.015		3.637	3.625	3.597		3.375	3.403		3.622	3.378		3.637		.036
238	3.484	.015		3.762	3.750	3.722		3.500	3.528		3.747	3.503		3.762		.038
239	3.609	.015		3.887	3.875	3.847		3.625	3.653		3.872	3.628		3.887		.039
240	3.734	.015		4.012	4.000	3.972		3.750	3.778		3.997	3.753		4.012		.040
241	3.859	.015		4.137	4.125	4.097		3.875	3.903		4.122	3.878		4.137		.041
242	3.984	.015		4.262	4.250	4.222		4.000	4.028		4.247	4.003		4.262		.043
243	4.109	.015		4.387	4.375	4.347		4.125	4.153		4.372	4.218		4.387		.044
244	4.234	.015		4.512	4.500	4.472		4.250	4.278		4.497	4.253		4.512		.045
245	4.359	.015		4.637	4.625	4.597		4.375	4.403		4.622	4.378		4.637		.046
246	4.484	.015		4.762	4.750	4.722		4.500	4.528		4.747	4.503		4.762		.048
247	4.609	.015		4.887	4.875	4.847		4.625	4.653		4.872	4.628		4.887		.049

Table 13.   (continued)

O-Ring Size Parker No. 2—	ID	±(1)	W	Mean OD (Ref)	A BORE DIA. (male gland) +.002 -.000	A-1 GROOVE DIA. (female gland) -.000	+	B TUBE OD (female gland) +.000 -.002	B-1 GROOVE DIA. (male gland) +.000	-	C PLUG DIA. (male gland) +.000 -.001	D THROAT DIA. (female gland) +.001 -.000	Gt GROOVE WIDTH	Ho OD FACE GROOVE (Use for Internal Pressure only) +.000	-	F DEPTH FACE GROOVE
248	4.734	.015		5.012	5.000	4.972		4.750	4.778		4.997	4.753		5.012	.050	
249	4.859	.015		5.137	5.125	5.097		4.875	4.903		5.122	4.878		5.137	.051	
250	4.984	.015		5.262	5.250	5.222		5.000	5.028		5.247	5.003		5.262	.053	
251	5.109	.023		5.387	5.375	5.347		5.125	5.153		5.372	5.128		5.387	.054	
252	5.234	.023		5.512	5.500	5.472		5.250	5.278		5.497	5.253		5.512	.055	
253	5.359	.023		5.637	5.625	5.597		5.375	5.403		5.622	5.378		5.637	.056	
254	5.484	.023		5.762	5.750	5.722		5.500	5.528		5.747	5.503		5.762	.057	
255	5.609	.023		5.887	5.875	5.847		5.625	5.653		5.872	5.628		5.887	.059	
256	5.734	.023		6.012	6.000	5.972		5.750	5.778		5.997	5.753		6.012	.060	
257	5.859	.023		6.137	6.125	6.097		5.875	5.903		6.122	5.878		6.137	.060	
258	5.984	.023		6.262	6.250	6.222		6.000	6.028		6.247	6.003		6.262	.060	
259	6.234	.023		6.512	6.500	6.472		6.250	6.278		6.497	6.253		6.512	.060	
260	6.484	.023		6.762	6.750	6.722		6.500	6.528		6.747	6.503		6.762	.060	
261	6.734	.023		7.012	7.000	6.972		6.750	6.778		6.997	6.753		7.012	.060	
262	6.984	.030		7.262	7.250	7.222		7.000	7.028		7.247	7.003		7.262	.060	
263	7.234	.030		7.512	7.500	7.472		7.250	7.278		7.497	7.253		7.512	.060	
264	7.484	.030		7.762	7.750	7.722		7.500	7.528		7.747	7.503		7.762	.060	
265	7.734	.030	.139	8.012	8.000	7.972	.002	7.750	7.778	.002	7.997	7.753	.187	8.012	.060	.111
266	7.984	.030	±.004	8.262	8.250	8.222		8.000	8.028		8.247	8.003	+.005	8.262	.060	+.002
267	8.234	.030		8.512	8.500	8.472		8.250	8.278		8.497	8.253	−.000	8.512	.060	−.000
268	8.484	.030		8.762	8.750	8.722		8.500	8.528		8.747	8.503		8.762	.060	
269	8.734	.030		9.012	9.000	8.972		8.750	8.778		8.997	8.753		9.012	.060	
270	8.984	.030		9.262	9.250	9.222		9.000	9.028		9.247	9.003		9.262	.060	
271	9.234	.030		9.512	9.500	9.472		9.250	9.278		9.497	9.253		9.512	.060	
272	9.484	.030		9.762	9.750	9.722		9.500	9.528		9.747	9.503		9.762	.060	
273	9.734	.030		10.012	10.000	9.972		9.750	9.778		9.997	9.753		10.012	.060	
274	9.984	.030		10.262	10.250	10.222		10.000	10.028		10.247	10.003		10.262	.060	
275	10.484	.030		10.762	10.750	10.722		10.500	10.528		10.747	10.503		10.762	.060	
276	10.984	.030		11.262	11.250	11.222		11.000	11.028		11.247	11.003		11.262	.060	
277	11.484	.030		11.762	11.750	11.722		11.500	11.528		11.747	11.503		11.762	.060	
278	11.984	.030		12.262	12.250	12.222		12.000	12.028		12.247	12.003		12.262	.060	
279	12.984	.030		13.262	13.250	13.222		13.000	13.028		13.247	13.003		13.262	.060	
280	13.984	.030		14.262	14.250	14.222		14.000	14.028		14.247	14.003		14.262	.060	
281	14.984	.030		15.262	15.250	15.222		15.000	15.028		15.247	15.003		15.262	.060	
282	15.955	.045		16.233	16.250	16.222		16.000	16.028		16.247	16.003		16.233	.060	
283	16.955	.045		17.233	17.250	17.222		17.000	17.028		17.247	16.003		17.233	.060	
284	17.955	.045		18.233	18.250	18.222		18.000	18.028		18.247	18.003		18.233	.060	
309	.412	.005		.832	.812	.777		.437	472		* .809	.440		.832	.008	
310	.475	.005	.210	.895	.875	.840		.500	.535		* .872	.503	.281	.895	.009	.170
311	.537	.005	±.005	.957	.937	.902	.004	.562	.597	.004	* .934	.565	+.005	.957	.010	+.003
312	.600	.005		1.020	1.000	.965		.625	.660		.997	.628	−.000	1.020	.010	−.000
313	.662	.005		1.082	1.062	1.027		.687	.722		1.059	.690		1.082	.011	
314	.725	.005		1.145	1.125	1.090		.750	.785		1.122	.753		1.145	.011	

Table 13. (continued)

O-Ring Size Parker No. 2-	ID	±(1)	W	Mean OD (Ref)	A BORE DIA. (male gland) +.002/-.000	A-1 GROOVE DIA. (female gland) -.000	+	B TUBE OD (female gland) +.000/-.002	B-1 GROOVE DIA. (male gland) +.000	-	C PLUG DIA. (male gland) +.000/-.001	D THROAT DIA. (female gland) +.001/-.000	G† GROOVE WIDTH	H₀ OD FACE GROOVE (Use for Internal Pressure only) +.000	F DEPTH FACE GROOVE -
315	.787	.006		1.207	1.187	1.152		.812	.847		1.184	.815		1.207	.012	
316	.850	.006		1.270	1.250	1.215		.875	.910		1.247	.878		1.270	.013	
317	.912	.006		1.332	1.312	1.277		.937	.972		1.309	.940		1.332	.013	
318	.975	.006		1.395	1.375	1.340		1.000	1.035		1.372	1.003		1.395	.014	
319	1.037	.006		1.457	1.437	1.402		1.062	1.097		1.434	1.065		1.457	.015	
320	1.100	.006		1.520	1.500	1.465		1.125	1.160		1.497	1.128		1.520	.015	
321	1.162	.006		1.582	1.562	1.527		1.187	1.222		1.559	1.190		1.582	.016	
322	1.225	.006		1.645	1.625	1.590		1.250	1.285		1.622	1.253		1.645	.016	
323	1.287	.006		1.707	1.687	1.652		1.312	1.347		1.684	1.315		1.707	.017	
324	1.350	.006		1.770	1.750	1.715		1.375	1.410		1.747	1.378		1.770	.018	
325	1.475	.010		1.895	1.875	1.840		1.500	1.535		1.872	1.503		1.895	.019	
326	1.600	.010		2.020	2.000	1.965		1.625	1.660		1.997	1.628		2.020	.020	
327	1.725	.010		2.145	2.125	2.090		1.750	1.785		2.122	1.753		2.145	.021	
328	1.850	.010		2.270	2.250	2.215		1.875	1.910		2.247	1.878		2.270	.023	
329	1.975	.010		2.395	2.375	2.340		2.000	2.035		2.372	2.003		2.395	.024	
330	2.100	.010		2.520	2.500	2.465		2.125	2.160		2.497	2.128		2.520	.025	
331	2.225	.010		2.645	2.625	2.590		2.250	2.285		2.622	2.253		2.645	.026	
332	2.350	.010		2.770	2.750	2.715		2.375	2.410		2.747	2.378		2.770	.028	
333	2.475	.010		2.895	2.875	2.840		2.500	2.535		2.872	2.503		2.895	.029	
334	2.600	.010		3.020	3.000	2.965		2.625	2.660		2.997	2.628		3.020	.030	
335	2.725	.015		3.145	3.125	3.090		2.750	2.785		3.122	2.753	.281	3.145	.031	.170
336	2.850	.015	.210	3.270	3.250	3.215	004	2.875	2.910	004	3.247	2.878	+.005	3.273	.033	+.003
337	2.975	.015	±.005	3.395	3.375	3.340		3.000	3.035		3.372	3.003	-.000	3.395	.034	-.000
338	3.100	.015		3.520	3.500	3.465		3.125	3.160		3.497	3.128		3.520	.035	
339	3.225	.015		3.645	3.625	3.590		3.250	3.285		3.622	3.253		3.645	.036	
340	3.350	.015		3.770	3.750	3.715		3.375	3.410		3.747	3.378		3.770	.038	
341	3.475	.015		3.895	3.875	3.840		3.500	3.535		3.872	3.502		3.895	.039	
342	3.600	.015		4.020	4.000	3.965		3.625	3.660		3.997	3.628		4.020	.040	
343	3.725	.015		4.145	4.125	4.090		3.750	3.785		4.122	3.753		4.145	.041	
344	3.850	.015		4.270	4.250	4.215		3.875	3.910		4.247	3.878		4.270	.043	
345	3.975	.015		4.395	4.375	4.340		4.000	4.035		4.372	4.003		4.395	.044	
346	4.100	.015		4.520	4.500	4.465		4.125	4.160		4.497	4.128		4.520	.045	
347	4.225	.015		4.645	4.625	4.590		4.250	4.285		4.622	4.253		4.645	.046	
348	4.350	.015		4.770	4.750	4.175		4.375	4.410		4.747	4.378		4.770	.047	
349	4.475	.015		4.895	4.875	4.840		4.500	4.535		4.872	4.503		4.895	.049	
350	4.600	.015		5.020	5.000	4.965		4.625	4.660		4.997	4.628		5.020	.050	
351	4.725	.015		5.145	5.125	5.090		4.750	4.785		5.122	4.753		5.145	.051	
352	4.850	.015		5.270	5.250	5.215		4.875	4.910		5.247	4.878		5.270	.053	
353	4.975	.015		5.395	5.375	5.340	-	5.000	5.035		5.372	5.003		5.395	.054	
354	5.100	.023		5.520	5.500	5.465		5.125	5.160		5.497	5.128		5.520	.055	
355	5.225	.023		5.645	5.625	5.590		5.250	5.285		5.622	5.253		5.645	.056	
356	5.350	.023		5.770	5.750	5.715		5.375	5.410		5.747	5.378		5.700	.057	
357	5.475	.023		5.895	5.875	5.840		5.500	5.535		5.872	5.503		5.895	.059	
358	5.600	.023		6.020	6.000	5.965		5.625	5.660		5.997	5.628		6.020	.060	

Table 13. (continued)

O-Ring Size Parker No. 2—	ID	±(1)	W	Mean OD (Ref)	A BORE DIA (male gland) +.002/−.000	A-1 GROOVE DIA (female gland) −.000	B TUBE OD (female gland) +.000/−.002	B-1 GROOVE DIA (male gland) +.000	C PLUG DIA (male gland) +.000/−.001	D THROAT DIA (female gland) +.001/−.000	G† GROOVE WIDTH	H₀ OD FACE GROOVE (Use for Internal Pressure only) +.000		F DEPTH FACE GROOVE
359	5.725	.023		6.145	6.125	6.090	5.750	5.785	6.122	5.753		6.145	.060	
360	5.850	.023		6.270	6.250	6.215	5.875	5.910	6.247	5.878		6.270	.060	
361	5.975	.023		6.395	6.375	6.340	6.000	6.035	6.372	6.003		6.395	.060	
362	6.225	.023		6.645	6.625	6.590	6.250	6.285	6.622	6.253		6.645	.060	
363	6.475	.023		6.895	6.875	6.840	6.500	6.535	6.872	6.503		6.895	.060	
364	6.725	.023		7.145	7.125	7.090	6.750	6.785	7.122	6.753		7.145	.060	
365	6.975	.023		7.395	7.375	7.340	7.000	7.035	7.372	7.003		7.395	.060	
366	7.225	.030		7.645	7.625	7.590	7.250	7.285	7.622	7.253		7.645	.060	
367	7.475	.030		7.895	7.875	7.840	7.500	7.535	7.872	7.503		7.895	.060	
368	7.725	.030		8.145	8.125	8.090	7.750	7.785	8.122	7.753		8.145	.060	
369	7.975	.030		8.395	8.375	8.340	8.000	8.035	8.372	8.003		8.395	.060	
370	8.225	.030		8.645	8.625	8.590	8.250	8.285	8.622	8.253		8.645	.060	
371	8.475	.030		8.895	8.875	8.840	8.500	8.535	8.872	8.503		8.895	.060	
372	8.725	.030		9.145	9.125	9.090	8.750	8.785	9.122	8.753		9.145	.060	
373	8.975	.030		9.395	9.375	9.340	9.000	9.035	9.372	9.003		9.395	.060	
374	9.225	.030	▼	9.645	9.625	9.590	9.250	9.285	9.622	9.253	▼	9.645	.060	▼
375	9.475	.030		9.895	9.875	9.840	9.500	9.535	9.872	9.503		9.895	.060	
376	9.725	.030		10.145	10.125	10.090	9.750	9.785	10.122	9.753	.281	10.145	.060	.170
377	9.975	.030	.210	10.395	13.375	10.340 .004	10.000	13.035 .004	10.372	10.003	+.005	10.395	.060	+.003
378	10.475	.030	±.005	10.895	10.875	10.840	10.500	10.535	10.872	10.503	−.000	10.895	.060	−.000
379	10.975	.030		11.395	11.375	11.340	11.000	11.035	11.372	11.003		11.395	.060	
380	11.475	.030		11.895	11.875	11.840	11.500	11.535	11.872	11.503		11.895	.060	
381	11.975	.030		12.395	12.375	12.340	12.000	12.035	12.372	12.003		12.395	.060	
382	12.975	.030		13.395	13.375	13.340	13.000	13.035	13.372	13.003		13.395	.060	
383	13.975	.030		14.395	14.375	14.340	14.000	14.035	14.372	14.003		14.395	.060	
384	14.975	.030		15.395	15.375	15.340	15.000	15.035	15.372	15.003		15.395	.060	
385	15.955	.045		16.375	16.375	16.340	16.000	16.035	16.372	16.003		16.375	.060	
386	16.955	.045		17.375	17.375	17.340	17.000	17.035	17.372	17.003		17.375	.060	
387	17.955	.045		18.375	18.375	18.340	18.000	18.035	18.372	18.003		18.375	.060	
388	18.953	.047		19.373	19.375	19.340	19.000	19.035	19.372	19.003		19.373	.060	
389	19.953	.047		20.373	20.375	20.340	20.000	20.035	20.372	20.003		20.373	.060	
390	20.953	.047		21.373	21.375	21.340	21.000	21.035	21.372	21.003		21.373	.060	
391	21.953	.047		22.373	22.375	22.340	22.000	22.035	22.372	22.003		22.373	.060	
392	22.940	.060		23.360	23.375	23.340	23.000	23.035	23.372	23.003		23.360	.060	
393	23.940	.060		24.360	24.375	24.340	24.000	24.035	24.372	24.003		24.360	.060	
394	24.940	.060		25.360	25.375	25.340	25.000	25.035	25.372	25.003		25.360	.060	
395	25.940	.060	▼	26.360	26.375	26.340	26.000	26.035	26.372	26.003	▼	26.360	.060	▼
425	4.475	.015	▲	5.025	5.000	4.952	4.500	4.548	4.996	4.504	▲	5.025	.050	▲
426	4.600	.015		5.150	5.125	5.077	4.625	4.673	5.121	4.629		5.150	.052	
427	4.725	.015	.275	5.275	5.250	5.202	4.750	4.798	5.246	4.754	.375	5.275	.053	.226
428	4.850	.015	±.006	5.400	5.375	5.327 .004	4.875	4.923 .004	5.371	4.879	+.005	5.400	.054	+.003
429	4.975	.015		5.525	5.500	5.452	5.000	5.048	5.496	5.004	−.000	5.525	.055	−.000
430	5.100	.023		5.650	5.625	5.577	5.125	5.173	5.621	5.129		5.650	.057	
431	5.225	.023	▼	5.775	5.750	5.702	5.250	5.298	5.746	5.254	▼	5.775	.058	▼

Table 13. (continued)

Parker No. 2- ID	±(1)	W	Mean OD (Ref)	A — BORE DIA. (male gland)	A-1 — GROOVE DIA. (female gland)	B — TUBE OD (female gland)	B-1 — GROOVE DIA. (male gland)	C — PLUG DIA. (male gland)	D — THROAT DIA. (female gland)	G† — GROOVE WIDTH	H₀ — OD FACE GROOVE (Use for Internal Pressure only)	F — DEPTH FACE GROOVE
(tolerance)				+.002 / −.000	+ / −.000	+.000 / −.002	+.000	+.000 / −.001	+.001 / −.000		+.000 / −	
432 5.350	.023		5.900	5.875	5.827	5.375	5.423	5.871	5.379		5.900	.059
433 5.475	.023		6.025	6.000	5.952	5.500	5.548	5.996	5.504		6.025	.060
434 5.600	.023		6.150	6.125	6.077	5.625	5.673	6.121	5.629		6.150	.060
435 5.725	.023		6.275	6.250	6.202	5.750	5.798	6.246	5.754		6.275	.060
436 5.850	.023		6.400	6.375	6.327	5.875	5.923	6.371	5.879		6.400	.060
437 5.975	.023		6.525	6.500	6.452	6.000	6.048	6.496	6.004		6.525	.060
438 6.225	.023		6.775	6.750	6.702	6.250	6.298	6.746	6.254		6.775	.060
439 6.475	.023		7.025	7.000	6.952	6.500	6.548	6.996	6.504		7.025	.060
440 6.725	.023		7.275	7.250	7.202	6.750	6.798	7.246	6.754		7.275	.060
441 6.975	.023		7.525	7.500	7.452	7.000	7.048	7.496	7.004		7.525	.060
442 7.225	.030		7.775	7.750	7.702	7.250	7.298	7.746	7.254		7.775	.060
443 7.475	.030		8.025	8.000	7.952	7.500	7.548	7.996	7.504		8.025	.060
444 7.725	.030		8.275	8.250	8.202	7.750	7.798	8.246	7.754		8.275	.060
445 7.975	.030		8.525	8.500	8.452	8.000	8.048	8.496	8.004		8.525	.060
446 8.475	.030		9.025	9.000	8.952	8.500	8.548	8.996	8.504		9.025	.060
447 8.975	.030		9.525	9.500	9.452	9.000	9.048	9.496	9.004		9.525	.060
448 9.475	.030		10.025	10.000	9.952	9.500	9.548	9.996	9.504		10.025	.060
449 9.975	.030		10.525	10.500	10.452	10.000	10.048	10.496	10.000		10.525	.060
450 10.475	.030		11.025	11.000	10.952	10.500	10.548	10.996	10.504		11.025	.060
451 10.975	.030		11.525	11.500	11.452	11.000	11.048	11.496	11.004		11.525	.060
452 11.475	.030		12.025	12.000	11.952	11.500	11.548	11.996	11.504		12.025	.060
453 11.975	.030	.275	12.525	12.500	12.452	12.000	12.048	12.496	12.004	.375	12.525	.060 .226
454 12.475	.030	±.006	13.025	13.000	12.952	12.500	12.548	12.996	12.504	+.005	13.025	.060 +.003
455 12.975	.030	−.000	13.525	13.500	13.452	13.000	13.048	13.496	13.004	−.000	13.525	.060 −.000
456 13.475	.030		14.025	14.000	13.952	13.500	13.548	13.996	13.504		14.025	.060
457 13.975	.030		14.525	14.500	14.452	14.000	14.048	14.496	14.004		14.525	.060
458 14.475	.030		15.025	15.000	14.952	14.500	14.548	14.996	14.504		15.025	.060
459 14.975	.030		15.525	15.500	15.452	15.000	15.048	15.496	15.004		15.525	.060
460 15.475	.030		16.025	16.000	15.952	15.500	15.548	15.996	15.504		16.025	.060
461 15.955	.045		16.505	16.500	16.452	16.000	16.048	16.496	16.004		16.505	.060
462 16.455	.045		17.005	17.000	16.952	16.500	16.548	16.996	16.504		17.005	.060
463 16.955	.045		17.505	17.500	17.452	17.000	17.048	17.496	17.004		17.505	.060
464 17.455	.045		18.005	18.000	17.952	17.500	17.548	17.996	17.504		18.005	.060
465 17.955	.045		18.505	18.500	18.452	18.000	18.048	18.496	18.004		18.505	.060
466 18.455	.045		19.005	19.000	18.952	18.500	18.548	18.996	18.504		19.005	.060
467 18.955	.045		19.505	19.500	19.452	19.000	19.048	19.496	19.004		19.505	.060
468 19.455	.045		20.025	20.000	19.952	19.500	19.548	19.996	19.504		20.005	.060
469 19.955	.045		20.505	20.500	20.452	20.000	20.048	20.496	20.004		20.505	.060
470 20.955	.045		21.505	21.500	21.452	21.000	21.048	21.496	21.004		21.505	.060
471 21.955	.045		22.505	22.500	22.452	22.000	22.048	22.496	22.004		22.505	.060
472 22.940	.060		23.490	23.500	23.452	23.000	23.048	23.496	23.004		23.490	.060
473 23.940	.060		24.490	24.500	24.452	24.000	24.048	24.496	24.004		24.490	.060
474 24.940	.060		25.490	25.500	25.452	25.000	25.048	25.496	25.004		25.490	.060
475 25.940	.060		26.940	26.500	26.452	26.000	26.048	26.496	26.004		26.490	.060

Note: For sizes 453–455, the A-1 and B-1 tolerances are +.004.

†This groove width does not permit the use of back-up rings. For pressures above 1500 psi, consult design chart for groove widths where back-up must be used.

*These designs require considerable installation stretch. If assembly breakage is incurred, use a compound having higher elongation or use a two-piece piston.

(1) Inside diameter tolerances for O-rings with inside diameters over 1/2 inch have been increased per AS568A. See Table 22, Chapter 6.

Basic Dimensions of "Parker 2-" O-rings are equivalent to AS568A.

Source: O-Ring Handbook OR5700, Parker Seal Co., Lexington, Ky., 1977.

Table 14. Industrial Specification for Reciprocating O-Ring Packing Glands

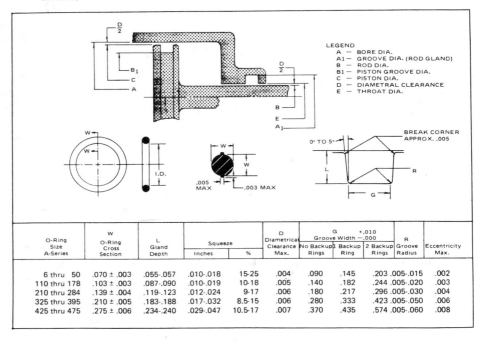

| O-Ring Size A-Series | W O-Ring Cross Section | L Gland Depth | Squeeze | | D Diametrical Clearance Max. | G Groove Width +.010 −.000 | | | R Groove Radius | Eccentricity Max. |
			Inches	%		No Backup Rings	1 Backup Ring	2 Backup Rings		
6 thru 50	.070 ± .003	.055-.057	.010-.018	15-25	.004	.090	.145	.203	.005-.015	.002
110 thru 178	.103 ± .003	.087-.090	.010-.019	10-18	.005	.140	.182	.244	.005-.020	.003
210 thru 284	.139 ± .004	.119-.123	.012-.024	9-17	.006	.180	.217	.296	.005-.030	.004
325 thru 395	.210 ± .005	.183-.188	.017-.032	8.5-15	.006	.280	.333	.423	.005-.050	.006
425 thru 475	.275 ± .006	.234-.240	.029-.047	10.5-17	.007	.370	.435	.574	.005-.060	.008

glands (see the "Design Chart"). Table 14 presents the piston and rod gland designs for the industrial reciprocating O-ring specification. The military and industrial specifications can be analyzed according to the squeeze and stretch they establish for the O-ring.

II. SPECIFIC DISCREPANCIES

A. Squeeze

O-ring squeeze is the percentage of cross-sectional compression experienced by an O-ring when installed in a particular gland design. The dimensions of the gland establish the O-ring squeeze. The military specification (Table 12) establishes an overall average O-ring squeeze of 15.5%:18.25% squeeze for 0.070-in. cross sections to 12.75% squeeze for 0.275-in. cross sections. The indus-

Table 14. (continued)

O-Ring Size AS568 Dash No.	DIMENSIONS				A BORE DIA +.002 −.000	A-1 GROOVE DIA (ROD) +.002 −.000	B ROD DIA +.000 −.002	B-1 GROOVE DIA (PIST.) +.000 −.002	C PISTON DIA +.000 −.001	E ROD BORE DIA +.001 −.000	G GLAND WIDTH +.005 −.000
	ID	±	W	Mean OD (REF)							
006	.114	.005		.254	.249	.234	.124	.139	.247	.126	
007	.145	.005		.285	.280	.265	.155	.170	.278	.157	
008	.176	.005	.070	.316	.311	.296	.186	.201	.309	.188	
009	.208	.005	±.003	.348	.343	.328	.218	.233	.341	.220	.093
010	.239	.005		.379	.374	.359	.249	.264	.372	.251	
011	.301	.005		.441	.436	.421	.311	.326	.434	.313	
012	.364	.005		.504	.499	.484	.374	.389	.497	.376	
104	.112	.005		.318	.312	.300	.124	.136	.310	.126	
105	.143	.005		.349	.343	.331	.155	.167	.341	.157	
106	.174	.005		.380	.374	.362	.186	.198	.372	.188	
107	.206	.005		.412	.406	.394	.218	.230	.404	.220	
108	.237	.005		.443	.437	.425	.249	.261	.435	.251	
109	.299	.005		.505	.499	.487	.311	.323	.497	.313	
110	.362	.005	.103	.568	.562	.550	.374	.386	.560	.376	.140
111	.424	.005	±.003	.630	.624	.612	.436	.448	.622	.438	
112	.487	.005		.693	.687	.675	.499	.511	.685	.501	
113	.549	.005		.755	.749	.737	.561	.573	.747	.563	
114	.612	.005		.818	.812	.800	.624	.636	.810	.626	
115	.674	.005		.880	.874	.862	.686	.698	.872	.688	
116	.737	.005		.943	.937	.925	.749	.761	.935	.751	
201	.171	.005		.449	.437	.427	.185	.195	.434	.188	
202	.234	.005		.512	.500	.490	.248	.258	.497	.251	
203	.296	.005		.574	.562	.552	.310	.320	.559	.313	
204	.359	.005		.637	.625	.615	.373	.383	.622	.376	
205	.421	.005		.699	.687	.677	.435	.445	.684	.438	
206	.484	.005		.762	.750	.740	.498	.508	.747	.501	
207	.546	.005		.824	.812	.802	.560	.570	.809	.563	
208	.609	.005		.887	.875	.865	.623	.633	.872	.626	
209	.671	.005		.949	.937	.927	.685	.695	.934	.688	
210	.734	.006		1.012	1.000	.990	.748	.758	.997	.751	
211	.796	.006	.139	1.074	1.062	1.052	.810	.820	1.059	.813	.187
212	.859	.006	±.004	1.137	1.125	1.115	.873	.883	1.122	.876	
213	.921	.006		1.199	1.187	1.177	.935	.945	1.184	.938	
214	.984	.006		1.262	1.250	1.240	.998	1.008	1.247	1.001	
215	1.046	.006		1.324	1.312	1.302	1.060	1.070	1.309	1.063	
216	1.109	.006		1.387	1.375	1.365	1.123	1.133	1.372	1.126	
217	1.171	.006		1.449	1.437	1.427	1.185	1.195	1.434	1.188	
218	1.234	.006		1.512	1.500	1.490	1.248	1.258	1.497	1.251	
219	1.296	.006		1.574	1.562	1.552	1.310	1.320	1.559	1.313	
220	1.359	.006		1.637	1.625	1.615	1.373	1.383	1.622	1.376	

trial specification for static O-ring applications (Table 13) establishes an overall average 20.5% O-ring squeeze: 27.0% squeeze for 0.70-in. cross sections to 18.0% squeeze for 0.275-in. cross sections. Hence the military specification requires less O-ring squeeze than that required by the industrial specification. Design Example 1 (see Sec. II.C) shows that this difference can be a factor of almost 2:1. The industrial specification for reciprocating gland designs (Table 14) indicates an overall average of 14.5% O-ring squeeze: 20% squeeze for 0.070-in. cross sections to

Table 14. (continued)

O-Ring Size AS568 Dash No.	DIMENSIONS				A BORE DIA +.002 −.000	A-1 GROOVE DIA (ROD) +.002 −.000	B ROD DIA +.000 −.002	B-1 GROOVE DIA (PIST.) +.000 −.002	C PISTON DIA +.000 −.001	E ROD BORE DIA +.001 −.000	G GLAND WIDTH +.005 −.000
	ID	±	W	Mean OD (REF)							
221	1.421	.006	.139	1.699	1.687	1.677	1.435	1.445	1.684	1.438	.187
222	1.484	.006	±.004	1.762	1.750	1.740	1.498	1.508	1.747	1.501	
309	.412	.005		.832	.812	.805	.435	.442	.809	.438	
310	.475	.005		.895	.875	.868	.498	.505	.872	.501	
311	.537	.005		.957	.937	.930	.560	.567	.934	.563	
312	.600	.005		1.020	1.000	.993	.623	.630	.997	.626	
313	.662	.005		1.082	1.062	1.055	.685	.692	1.059	.688	
314	.725	.005		1.145	1.125	1.118	.748	.755	1.122	.751	
315	.787	.006		1.207	1.187	1.180	.810	.817	1.184	.813	
316	.850	.006		1.270	1.250	1.243	.873	.880	1.247	.876	
317	.912	.006		1.332	1.312	1.305	.935	.942	1.309	.938	
318	.975	.006		1.395	1.375	1.368	.998	1.005	1.372	1.001	
319	1.037	.006		1.457	1.437	1.430	1.060	1.067	1.434	1.063	
320	1.100	.006		1.520	1.500	1.493	1.123	1.130	1.497	1.126	
321	1.162	.006		1.582	1.562	1.555	1.185	1.192	1.559	1.188	
322	1.225	.006		1.645	1.625	1.618	1.248	1.255	1.622	1.251	
323	1.287	.006		1.707	1.687	1.680	1.310	1.317	1.684	1.313	
324	1.350	.006		1.770	1.750	1.743 +.004 −.000	1.373	1.380 +.000 −.004	1.747	1.376	
325	1.475	.010		1.895	1.875	1.868	1.496	1.505	1.872	1.501	
326	1.600	.010		2.020	2.000	1.993	1.623	1.630	1.997	1.626	
327	1.725	.010	.210	2.145	2.125	2.118	1.748	1.755	2.122	1.751	.281
328	1.850	.010	±.005	2.270	2.250	2.243	1.873	1.880	2.247	1.876	
329	1.975	.010		2.395	2.375	2.368	1.998	2.005	2.372	2.001	
330	2.100	.010		2.520	2.500	2.493	2.123	2.130	2.497	2.126	
331	2.225	.010		2.645	2.625	2.618	2.248	2.255	2.622	2.251	
332	2.350	.010		2.770	2.750	2.743	2.373	2.380	2.747	2.376	
333	2.475	.010		2.895	2.875	2.868	2.498	2.505	2.872	2.501	
334	2.600	.010		3.020	3.000	2.993	2.623	2.630	2.997	2.626	
335	2.725	.010		3.145	3.125	3.118	2.748	2.755	3.122	2.751	
336	2.850	.015		3.270	3.250	3.243	2.873	2.880	3.247	2.876	
337	2.975	.015		3.395	3.375	3.368	2.998	3.005	3.372	3.001	
338	3.100	.015		3.520	3.500	3.493	3.123	3.130	3.497	3.126	
339	3.225	.015		3.645	3.625	3.618	3.248	3.255	3.622	3.251	
340	3.350	.015		3.770	3.750	3.743	3.373	3.380	3.747	3.376	
341	3.475	.015		3.895	3.875	3.868	3.498	3.505	3.872	3.501	
342	3.600	.015		4.020	4.000	3.993	3.623	3.630	3.997	3.626	
343	3.725	.015		4.145	4.125	4.118	3.748	3.755	4.122	3.751	
344	3.850	.015		4.270	4.250	4.243	3.873	3.880	4.247	3.876	

13.8% squeeze for 0.275-in. cross sections. Therefore, for reciprocating gland applications, the industrial and military specifications are basically the same, except that the industrial reciprocating gland specification continues for 26 additional O-ring sizes, giving more applications for the intermediate sizes of 0.103-, 0.139-, and 0.210-in. cross-sectional diameters.

For static gland design requiring large O-ring squeeze, especially in cases involving low temperatures or large temperature

Table 14.   (continued)

O-Ring Size AS568 Dash No.	DIMENSIONS				A BORE DIA +.002 -.000	A-1 GROOVE DIA (ROD) +.004 -.000	B ROD DIA +.000 -.002	B-1 GROOVE DIA (PIST.) +.000 -.004	C PISTON DIA +.000 -.001	E ROD BORE DIA +.001 -.000	G GLAND WIDTH +.005 -.000
	ID	±	W	Mean OD (REF)							
345	3.975	.015		4.395	4.375	4.368	3.998	4.005	4.372	4.001	
346	4.100	.015	.210	4.520	4.500	4.493	4.123	4.130	4.497	4.126	.281
347	4.225	.015	±.005	4.645	4.625	4.618	4.248	4.255	4.622	4.251	
348	4.350	.015		4.770	4.750	4.743	4.373	4.380	4.747	4.376	
349	4.475	.015		4.895	4.875	4.868	4.498	4.505	4.872	4.501	
425	4.475	.015		5.025	5.002	4.971	4.497	4.528	4.998	4.501	
426	4.600	.015		5.150	5.127	5.096	4.622	4.653	5.123	4.626	
427	4.725	.015		5.275	5.252	5.221	4.747	4.778	5.248	4.751	
428	4.850	.015		5.400	5.377	5.346	4.872	4.903	5.373	4.876	
429	4.975	.015		5.525	5.502	5.471	4.997	5.028	5.498	5.001	
430	5.100	.023		5.650	5.627	5.596	5.122	5.153	5.623	5.126	
431	5.225	.023		5.775	5.752	5.721	5.247	5.278	5.748	5.251	
432	5.350	.023		5.900	5.877	5.846	5.372	5.403	5.873	5.376	
433	5.475	.023		6.025	6.002	5.971	5.497	5.528	5.998	5.501	
434	5.600	.023		6.150	6.127	6.096	5.622	5.653	6.123	5.626	
435	5.725	.023		6.275	6.252	6.221	5.747	5.778	6.248	5.751	
436	5.850	.023		6.400	6.377	6.346	5.872	5.903	6.373	5.876	
437	5.975	.023		6.525	6.502	6.471	5.997	6.028	6.498	6.001	
438	6.225	.023		6.775	6.752	6.721	6.247	6.278	6.748	6.251	
439	6.475	.023		7.025	7.002	6.971	6.497	6.528	6.998	6.501	
440	6.725	.023		7.275	7.252	7.221	6.747	6.778	7.248	6.751	
441	6.975	.023		7.525	7.502	7.471	6.997	7.028	7.498	7.001	
442	7.225	.030	.275	7.775	7.752	7.721	7.247	7.278	7.748	7.251	.375
443	7.475	.030	±.006	8.025	8.002	7.971	7.497	7.528	7.998	7.501	
444	7.725	.030		8.275	8.252	8.221	7.747	7.778	8.248	7.751	
445	7.975	.030		8.525	8.502	8.471	7.997	8.028	8.498	8.001	
446	8.475	.030		9.025	9.002	8.971	8.497	8.528	8.998	8.501	
447	8.975	.030		9.525	9.502	9.471	8.997	9.028	9.498	9.001	
448	9.475	.030		10.025	10.002	9.971	9.497	9.528	9.998	9.501	
449	9.975	.030		10.525	10.502	10.471	9.997	10.028	10.498	10.001	
450	10.475	.030		11.025	11.002	10.971	10.497	10.528	10.998	10.501	
451	10.975	.030		11.525	11.502	11.471	10.997	11.028	11.498	11.001	
452	11.475	.030		12.025	12.002	11.971	11.497	11.528	11.998	11.501	
453	11.975	.030		12.525	12.502	12.471	11.997	12.028	12.498	12.001	
454	12.475	.030		13.025	13.002	12.971	12.497	12.528	12.998	12.501	
455	12.975	.030		13.525	13.502	13.471	13.497	13.028	13.498	13.001	
456	13.475	.030		14.025	14.002	13.971	13.497	13.528	13.998	13.501	
457	13.975	.030		14.525	14.502	14.471	13.997	14.028	14.498	14.001	
458	14.475	.030		15.025	15.002	14.971	14.497	14.528	14.998	14.501	
459	14.975	.030		15.525	15.502	15.471	14.997	15.028	15.498	15.001	
460	15.475	.030		16.025	16.002	15.971	15.497	15.528	15,998	15,501	

Source: *O-Ring Design and Selection Handbook 110-A,* Sargent
Industries, Carlsbad, Calif., 1977.

fluctuations, the industrial specification (Table 13) should be
used. Use the military specification (Table 12) when less O-ring
squeeze is desired, that is, when the design incorporates more
than two or three O-ring interfaces on parts that must be assembled with ease and when temperature fluctuations will not be
encountered.

B. Stretch

O-ring stretch is the percent increase in inside diameter (ID) of
an O-ring when installed in a male gland. The dimensions of the
male gland establishes the new, as installed, ID of the O-ring
(B-1 in Table 13). The industrial specification (Table 13) establishes more stretch on the inside diameter of the O-ring than
the military specification (Table 12). The O-ring diametral stretch
for the male gland according to the industrial specification varies
from 73 to 0.1%, the larger stretch being for the smaller O-ring
sizes. The military design specifies 14 to 0.15% O-ring diametral
stretch ([(0.033 − 0.029)/0.029] X 100% = 14%, for MS 28775, size
001 O-ring, Table 12).

When diametral stretch is greater than 5%, the stress produced
in the O-ring may cause a reduction in physical capabilities and
general life of the O-ring. The military specification requires a
5% and greater diametral stretch for O-ring sizes 008 and smaller
([(0.123 − 0.114)/0.114] X 100% = 18%, for the size 006 O-ring,
Table 12), whereas the industrial specification requires a 5% and
greater diametral stretch for O-ring sizes 116 and smaller
([0.650 − 0.612) / 0.612] X 100% = 6.2% for the size 114 O-ring,
Table 13). The designer should consider the reduction in cross-
sectional diameter of an O-ring when stretched beyond 5%. This
is important in accurately determing the squeeze for a small O-
ring, 1.00 in. ID and smaller. For most O-ring compounds, the
percent reduction in cross-sectional diameter of an O-ring is approximately one-half the percent of diametral stretch of the
O-ring. Hence an O-ring stretched 5% in a male gland has its
cross-sectional diameter reduced by 2.5%; one stretched 10% has
its cross-sectional diameter reduced by 5% (see Fig. 6 for reference).

C. Design Examples

Design Examples 1 and 2 show the difference in O-ring squeeze
and stretch as specified by the military and industrial specifications. Note that the percent O-ring stretch required by the industrial specification (Table 13) is about three times that required by the military specification (Table 12). This results in a
small O-ring cross section. Also note that the cross-sectional
squeeze may be twice as great for the O-ring when installed in
the industrial gland. This becomes particularly significant when
assembling parts, since O-ring squeeze determines compression
load. Figure 8 shows that compression loads are exponentially
dependent on the squeeze of an O-ring (see Chap. 4). There-

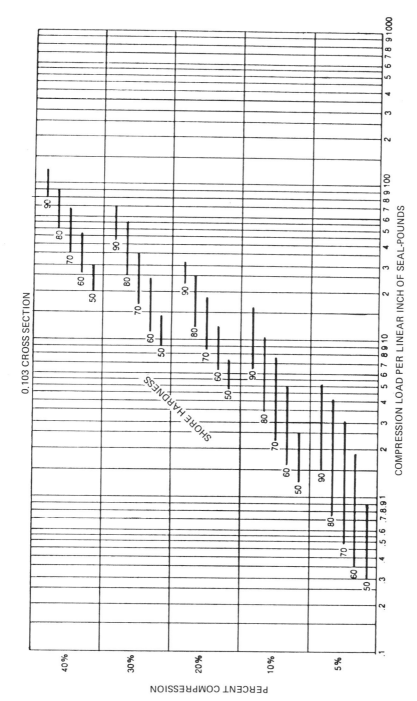

Figure 8. Compression load per linear inch of seal; pounds; (From Parker Hannifin Corp., Cleveland, Ohio; reproduced with permission)

fore, in Design Example 2 we find it will take three times the compression load (hence, the assembly force) to install the O-ring in the industrial-designed gland than in the military-designed gland.

Design Example 1 Comparison: Military Versus Industrial Specifications for O-Ring Size 342

ID = 3.600 ± 0.015

w = 0.210 ± 0.005

Military specification (MIL-G-5514), Table 12

$3.993 \begin{Bmatrix} + 0.002 \\ - 0.000 \end{Bmatrix}$ diam. $3.621 \begin{Bmatrix} + 0.000 \\ - 0.002 \end{Bmatrix}$ diam.

Percent O-ring stretch:

$$\frac{3.621 - 3.600}{3.600} \times 100\% = 0.58\%$$

O-ring cross section:

$$(0.210 \pm 0.005) \left(100\% - \frac{0.58\%}{2} \right) = 0.209 \pm 0.005 \text{ diameter}$$

Nominal percent cross-sectional squeeze:

$$\frac{0.209 - (3.993 - 3.621)/2}{0.209} \times 100\% = 11.0\%$$

Industrial specification, Table 13

$4.000 \begin{Bmatrix} + 0.002 \\ - 0.000 \end{Bmatrix}$ diam. $3.660 \begin{Bmatrix} + 0.000 \\ - 0.004 \end{Bmatrix}$ diam.

Percent O-ring stretch:

$$\frac{3.660 - 3.600}{3.600} \times 100\% = 1.67\%$$

O-ring cross section:

$$(0.210 \pm 0.005) \left(100\% - \frac{1.67\%}{2}\right) = 0.208 \pm 0.005 \text{ diameter}$$

Nominal percent cross-sectional squeeze:

$$\frac{0.208 - (4.000 - 3.660)/2}{0.208} \times 100\% = 18.3\%$$

Design Example 2. Comparison: Military Versus Industrial Specifications for O-Ring Size 116

    ID = 0.737 ± 0.005

    w = 0.103 ± 0.003

Military specification (MIL-G-5514), Table 12

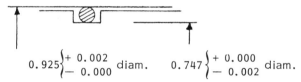

$$0.925 \begin{Bmatrix} + \ 0.002 \\ - \ 0.000 \end{Bmatrix} \text{ diam.} \qquad 0.747 \begin{Bmatrix} + \ 0.000 \\ - \ 0.002 \end{Bmatrix} \text{ diam.}$$

Percent O-ring stretch:

$$\frac{\left(0.747 \begin{Bmatrix} + \ 0.000 \\ - \ 0.002 \end{Bmatrix}\right) - (0.737 \pm 0.005)}{0.737 \pm 0.005} \times 100\% = \frac{0.003 \text{ to } 0.005}{0.732 \text{ to } 0.742} \times 100\%$$

$$\frac{0.003}{0.742} \times 100\% = 0.4\% \text{ minimum stretch}$$

$$\frac{0.005}{0.732} \times 100\% = 2.0\% \text{ maximum stretch}$$

O-ring cross section due to the maximum stretch:

$$(0.103 \pm 0.003) \left(100\% - \frac{2.0\%}{2}\right) = 0.102 \pm 0.003 \text{ diameter}$$

Percent cross-sectional squeeze:

$$\frac{(0.102 \pm 0.003) - \left[\left(0.925 \begin{smallmatrix} + \ 0.002 \\ - \ 0.000 \end{smallmatrix}\right) - \left(0.747 \begin{smallmatrix} + \ 0.000 \\ - \ 0.002 \end{smallmatrix}\right)\right]/2}{0.102 \pm 0.003} \times 100\%$$

$$= \frac{0.013 \begin{Bmatrix} + \ 0.003 \\ - \ 0.005 \end{Bmatrix}}{0.102 \pm 0.003} \times 100\% = 12.75\% \text{ nominal } (8.1\% \text{ minimum,}$$
$$15.2\% \text{ maximum})$$

Compression load on O-ring (according to Fig. 8, for a nominal 12.75% compression and 70 Shore hardness):

$$\frac{3 \text{ lb compression}}{\text{linear inch}} \times 0.925 \ \pi \text{ in.} = 8.7 \text{ lb}$$

Industrial specification, Table 13

$$0.937 \begin{Bmatrix} + 0.002 \\ - 0.000 \end{Bmatrix} \text{diam.} \qquad 0.775 \begin{Bmatrix} + 0.000 \\ - 0.002 \end{Bmatrix} \text{diam.}$$

Percent O-ring stretch:

$$\frac{\left(0.775 \begin{Bmatrix} + 0.000 \\ - 0.002 \end{Bmatrix}\right) - (0.737 \pm 0.005)}{0.737 \pm 0.005} \times 100\% = \frac{0.031 \text{ to } 0.043}{0.732 \text{ to } 0.742} \times 100\%$$

$$\frac{0.031}{0.742} \times 100\% = 4.2\% \text{ minimum stretch}$$

$$\frac{0.043}{0.732} \times 100\% = 5.9\% \text{ maximum stretch}$$

O-ring cross section due to the maximum stretch:

$$(0.103 \pm 0.003) \left(100\% - \frac{5.9\%}{2}\right) = 0.100 \pm 0.003 \text{ diameter}$$

Percent cross-sectional squeeze:

$$\frac{(0.100 \pm 0.003) - \left[\left(0.937 \begin{Bmatrix} + 0.002 \\ - 0.000 \end{Bmatrix}\right) - \left(0.775 \begin{Bmatrix} + 0.000 \\ - 0.002 \end{Bmatrix}\right)\right] \Big/ 2}{0.100 \pm 0.003} \times 100\%$$

$$= \frac{0.019 \begin{Bmatrix} + 0.003 \\ - 0.005 \end{Bmatrix}}{0.100 \pm 0.003} \times 100\% = 19.0\% \text{ nominal } (13.6\% \text{ minimun,}$$

$$21.4\% \text{ maximum})$$

Compression load on O-ring (according to Fig. 8, for a nominal 19.0% compression and 70 Shore hardness):

$$\frac{9 \text{ lb compression}}{\text{linear inch}} \ 0.937 \pi \text{ in} = 26.5 \text{ lb}$$

4

Static Seal Applications

Static seal applications are by far the major use of elastomeric
O-ring seals. Most static seals are employed as face seals, but
male and female radial seals are becoming more common. This
author prefers radial squeeze applications over axial squeeze face-
seal applications for two reasons: (1) radial seals depend on the
dimensions of the parts to maintain O-ring squeeze, whereas face
seals depend on external fasteners such as screws and bolts to
maintain O-ring squeeze; and (2) radial seals are usually easier
to employ when modifying existing parts, whereas face seals usually require additional radial space and flange protrusions.
These and other design considerations follow, but the information
presented in Chap. 3, especially Tables 13 and 14, should also be
consulted before attempting to design a static seal.

I. SQUEEZE

Industrial specifications are designed for an overall average 20.5%
O-ring squeeze: 27% squeeze for 0.070-in. cross sections to 18%
squeeze for 0.275-in. cross sections. The military specification
requires approximately one-half the O-ring squeeze specified by
the industrial specification, because it incorporates dynamic applications together with static applications. For reciprocating
glands, the industrial and military specifications are basically the
same. Use the industrial specification for gland designs requiring
greater O-ring squeeze, especially in cases involving low temperatures. Use the military specification for easier assembly of
parts that incorporate many O-ring interfaces and where temperature fluctuations are not encountered.

II. STRETCH

The industrial specification requires more stretch on the inside diameter of the O-ring than that required by the military specification. The O-ring diametral stretch for the male gland according to the industrial specification varies from 73 to 0.1%, the larger stretch for the smaller O-ring sizes. The military design specifies 14 to 0.15% O-ring diametral stretch. The designer should consider diametral stretch when dealing with O-ring size AS 568-116 and smaller. The designer may want to consider the reduction in cross-sectional diameter of an O-ring stretched beyond 5%. This relationship is given in Fig. 6. For applications requiring high diametral stretch, ethylene propylene, fluorocarbon, neoprene, or polyurethane O-rings are recommended.

In the male gland applications of 8 in. and larger, the design stretch is so small (0.1% diametral stretch) that the O-ring may tend to sag out of the groove. This makes it difficult to assemble the parts so as not to damage the O-ring. The problem can be solved by using the next size smaller O-ring in the larger male gland. The reduction in cross-sectional diameter of the O-ring must then be checked according to Fig. 6, and the gland depth may need to be reduced to maintain O-ring squeeze. Design Example 3 considers this case.

Design Example 3 Static Elastomeric Seal Application: Male Gland Involving Low Temperature

Given:

Male gland of approximately 11.75 in. piston diameter and nitrile O-ring to experience temperatures of 0°F maximum to −40°F minimum and an average differential pressure of 1800 psi.

Find:

The groove diameter (B-1) such that the ID stretch of the O-ring will not exceed 5%, and the bore diameter (A) will provide

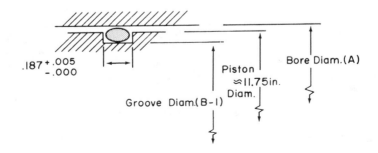

a nominal 15% O-ring squeeze without O-ring extrusion.
Solution:

1. Because nitrile is one of the worst materials for maintaining its physical properties in a stretched condition, the groove diameter (B-1) and O-ring size must be determined such that the ID stretch will not exceed 5% at the minimum temperature condition of −40°F. According to the table for industrial static seals (Table 13), an AS568-277 size O-ring should be used for a piston diameter of 11.75 and a groove width of 0.187 in. Because the nominal stretch for these larger O-rings is less than 0.1% and we do not want the O-ring to sag out of the groove during installation, we shall choose the next smaller O-ring, AS568-276. The ID of this O-ring is 10.984 ± 0.030 in. diameter, and in an environment of −40°F this diameter will decrease by

$$(10.984 \pm 0.030 \text{ in.})(6.2 \times 10^{-5} \text{ in./in. °F})[70°F - (-40°F]$$

$$= 0.0749 \pm 0.0002$$

or approximately 0.075 in. The O-ring ID will then be 10.909 in., and for 5% stretch the groove diameter must be

$$10.909 \text{ in. } (1.05) = 11.454 \begin{cases} + 0.000 \text{ in.} \\ - 0.002 \text{ in.} \end{cases}$$

= groove diam. (B-1) (Answer)

2. To provide an O-ring squeeze of 15%, we first refer to Fig. 6 for 5% ID stretch. The O-ring cross section will decrease by 3.5%, or

$$(0.139 \pm 0.004 \text{ in.})(0.035) = 0.0049 \text{ in.}$$

At −40°F the O-ring cross section will have decreased by

$$(0.139 \text{ in.})(6.2 \times 10^{-5} \text{ in./in. °F})(110°F) = 0.0009 \text{ in.}$$

Therefore, the O-ring cross section will be

$$(0.139 \pm 0.004 \text{ in.}) - 0.0049 \text{ in.} - 0.0009 \text{ in.} = 0.133 \pm 0.004 \text{ in.}$$

For 15% O-ring cross-sectional squeeze, the bore diameter (A) must be

$$11.454 \text{ in.} + 2(0.133 \text{ in.})(1.00 - 0.15) = 11.454 + 0.226$$

$$= 11.680 \text{ in.}$$

Rounding off to a standard bore diameter, we obtain

$$11.687 \begin{cases} + 0.002 \text{ in.} \\ - 0.000 \text{ in.} \end{cases} = \text{bore diam. (A) (Answer)}$$

Check on actual squeeze:

$$11.687 \begin{cases} + 0.002 \\ - 0.000 \end{cases} \text{bore diam.}$$

$$-11.454 \begin{cases} + 0.000 \\ - 0.002 \end{cases} \text{groove diam.}$$

$$\frac{0.233 \begin{cases} + 0.004 \\ - 0.000 \end{cases} = 0.1165 \begin{cases} + 0.002 \\ - 0.000 \end{cases} = 0.117 \begin{array}{c} + 0.002 \\ - 0.000 \end{array}}{2}$$

Actual squeeze:

$$0.133 \begin{cases} + 0.004 \\ - 0.004 \end{cases} \text{O-ring cross section before installed}$$

$$-0.117 \begin{cases} + 0.002 \\ - 0.000 \end{cases} \text{O-ring cross section installed}$$

$$0.016 \begin{cases} + 0.004 \\ - 0.006 \end{cases} = 0.010 \text{ min. to } 0.020 \text{ max.}$$

$$\frac{0.010 \text{ min.}}{0.133} \times 100\% = 7.5\% \text{ min. squeeze}$$

$$\frac{0.020 \text{ max.}}{0.133} \times 100\% = 15\% \text{ max. squeeze}$$

3. To make sure that the O-ring does not extrude under 1800 psi differential pressure, we consult Fig. 5. At 1800 psi, we find the maximum allowable gap to be about 0.002 in. for a 70-durometer O-ring. Therefore, the piston diameter should be specified as

$$(11.687 + 0.002 \text{ in.}) - 2(0.002) = 11.685 \text{ in. min.}$$

Note that the calculation considers the diametral gap to be 0.004; thus the piston must be held concentric to the cylinder bore.

or

$$11.686 \begin{cases} + 0.000 \\ - 0.001 \end{cases} \text{in. piston diam. (Answer)}$$

The actual gap could then range as follows:

$$11.687 \begin{cases} + 0.002 \\ - 0.000 \end{cases} \text{bore diam}$$

$$-11.686 \begin{cases} + 0.000 \\ - 0.001 \end{cases} \text{piston diam.}$$

$$\frac{0.001 \begin{cases} + 0.003 \\ - 0.000 \end{cases}}{2} = 0.0005 \text{ to } 0.0020 \text{ in.}$$

III. BACKUP RINGS

Both the industrial and military specifications recommend the use of backup rings for pressures greater than 1500 psi or when large pressure fluctuations are encountered. If parts are designed such that clearance gaps are zero or at least less than that specified in Fig. 5, backup rings need not be used. Pressure fluctuations or vibration of mating parts may produce a "pumping" phenomenon of the O-ring, causing slight leakage and eventual O-ring failure. In these cases, and especially for silicone O-rings, backup rings are recommended. Due to the poor antiextrusion capability of silicone, most O-ring manufacturers recommend reducing the gap clearances given in the design tables by 50% in applications involving silicone O-rings.

IV. SURFACE FINISH

A surface finish of 63 rms is recommended for static seal applications. This applies to the surfaces of which the O-ring will contact, including the sides of the gland. Some manufacturers recommend 63 rms finish on just the sides of the gland and 32 rms finish on the bottom surface of the gland and O-ring contact surface of the mating part. Parker Seal Company recommends a maximum 16 rms surface finish for face seals in gas and vacuum applications. Machining procedures for producing O-ring sealing surfaces must not produce scratches, ridges, or other discontinuities across the O-ring contact line. Such discontinuities parallel to the O-ring contact line may be acceptable, but are still undesirable.

V. STANDARD BOSSES AND FITTINGS

The design tables for fittings and bosses are presented under military standards MS16142, MS33649, and MS33656—Tables 15, 16, and 17, respectively. The two standards for the internal straight-thread boss (Tables 15 and 16) make up the female portion of a tube fitting gland. The tube fitting end (Table 17) can be used with either of the two boss configurations shown in Fig. 9. The boss configuration according to MS16142 (Table 15) has been adopted by general industry and is usually referred to as the industrial standard for tube fittings. The military usually incorporates MS33649 (Table 16), although there has been a recent trend to use MS16142 in more military applications. As can be seen in Fig. 9, the O-ring is subjected to a more severe mode of squeeze under the military standard, 38% cross-sectional squeeze compared to 26% cross-sectional squeeze for the industrial standard. The military standard is more apt to result in cutting the

Table 15. Boss, Straight-Thread Tube Fitting (Industrial-Adopted Specification MS16142)

Source: U.S. Government Printing Office, 603-108/774 (1974).

Table 16. Boss, Internal Straight Thread (MS33649)

DASH NUMBER	TUBE OD MM	THREAD T PER MIL-S-8879	A DIA +.015 -.000	B MIN SEE NOTE 7	C DIA	D DIA +.005 -.000	E +.015 -.000	G DIA MIN	J MIN	N
01	—	.2500-28 UNJF-3B	.359	.330	.062	.264		.478	.402	.002
02	.125	.3125-24 UNJF-3B	.438	.482		.328	.063	.602	.577	
03	.188	.3750-24 UNJF-3B	.500	.538	.125	.390		.665	.583	.003
04	.250	.4375-20 UNJF-3B	.562	.568	.172	.454	.075	.728	.656	
05	.312	.5000-20 UNJF-3B	.625		.234	.517		.790		
06	.375	.5625-18 UNJF-3B	.688	.598	.297	.580	.083	.852	.709	.004
07	.438	.6250-18 UNJF-3B	.750	.614	.360	.643	.094	.915	.725	
08	.500	.7500-16 UNJF-3B	.875	.714	.391	.769		1.040	.834	
09	.562	.8125-16 UNJ-3B	.938	.730	.438	.832	.107	1.102	.850	
10	.625	.8750-14 UNJF-3B	1.000	.802	.484	.896		1.165	.930	.005
11	.688	1.0000-12 UNJF-3B	1.156		.547	1.023		1.352		
12	.750	1.0625-12 UNJ-3B	1.234		.609	1.086		1.415	1.064	
14	.875	1.1875-12 UNJ-3B	1.362		.734	1.211		1.540		
16	1.000	1.3125-12 UNJ-3B	1.487	.877	.844	1.336		1.665		
18	1.125	1.5000-12 UNJ-3B	1.675		.953	1.524	.125	1.790	1.116	.008
20	1.250	1.6250-12 UNJ-3B	1.800		1.078	1.648		1.978		
24	1.500	1.8750-12 UNJ-3B	2.050		1.312	1.898		2.228	1.127	
28	1.750	2.2500-12 UNJ-3B	2.425		1.547	2.273		2.602	1.243	.010
32	2.000	2.5000-12 UNJ-3B	2.675	.907	1.781	2.524		2.852	1.368	

1. DIMENSIONS IN INCHES.

2. SURFACE TEXTURE: ANSI B46.1-1962.

3. REMOVE ALL BURRS AND SLIVERS AND BREAK SHARP EDGES.

4. MINIMUM DIAMETER FLAT SURFACE. CLEARANCE PROVISIONS FOR FITTING, WRENCH, FITTING INSTALLATION AND TOOL FILLET RADII MUST BE ADDED AS REQUIRED. G MIN DIAMETER MAY BE REDUCED BY .040 WHEN IT DEFINES THE LIMITS OF A RAISED SURFACE AND WHEN STRENGTH REQUIREMENTS PERMIT.

5. CERTAIN PROVISIONS OF THIS STANDARD ARE THE SUBJECT OF INTERNATIONAL STANDARDIZATION AGREEMENT ASCC AIR STD 17/18. WHEN REVISION OR CANCELLATION OF THIS STANDARD IS PROPOSED, THE DEPARTMENTAL CUSTODIANS WILL INFORM THEIR RESPECTIVE STANDARDIZATION OFFICE SO THAT APPROPRIATE ACTION MAY BE TAKEN RESPECTING THE INTERNATIONAL AGREEMENT CONCERNED.

6. THRU THREAD DESIGN SHALL HAVE SUFFICIENT THREADS TO MEET STRENGTH REQUIREMENTS OF MATERIAL USED.

7. B MIN IS THE FULL THREAD DEPTH REQUIREMENT FOR BLIND TAP DESIGN.

8. REFERENCE CLEARANCE ENVELOPE PROVIDES MINIMUM CLEARANCE FOR THE LONGEST STANDARD FITTING.

THIS STANDARD WAS DEVELOPED COOPERATIVELY WITH THE MILITARY SERVICES BY THE SAE AEROSPACE PROPULSION DIVISION AND PART STANDARDS DIVISION.

THIS IS A DESIGN STANDARD, NOT TO BE USED AS A PART NUMBER.

P.A. Air Force - 11 Other Cust Army - AV Navy - AS	INTERNATIONAL INTEREST	TITLE BOSSES, FLUID CONNECTION - INTERNAL STRAIGHT THREAD	MILITARY STANDARD MS 33649
PROCUREMENT SPECIFICATION NONE	SUPERSEDES:	AND 10049, AND 10050	SHEET 1 OF 1

Source: U.S. Government Printing Office, 703-020/4848 (1977).

Table 17. Flared Tube Fittings (MS 33656)

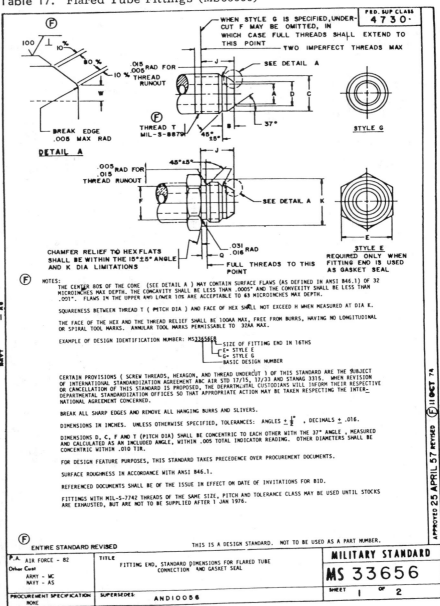

Table 17. (continued)

4730

TABLE I DIMENSIONS

SIZE NO	TUBING OD	THREAD (F) MIL-S-8879	A DIA	B +.015 -.000	C +.000 -.005	D +.003 DIA	E	F +.002 -.003	H MAX	J ±.015	K +.010 DIA	O +.015 -.000	X MIN
2	1/8	5/16-24UNJF-3A	.082	.177	.245	.083	.563	.250		.448	.549		
3	3/16	3/8-24UNJF-3A	.125		.307	.146	.625	.312		.479	.611	.063	
4	1/4	7/16-20UNJF-3A	.172	.193	.359	.193	.688	.364			.674	.075	.035
5	5/16	1/2-20UNJF-3A	.234		.421	.255	.750	.426		.550	.736		
6	3/8	9/16-18UNJF-3A	.297	.198	.476	.318	.813	.481	.005	.556	.799	.083	
8	1/2	3/4-16UNJF-3A	.391	.253	.654	.426	1.000	.660		.657	.986	.094	.012
10	5/8	7/8-14UNJF-3A	.484	.266	.767	.539	1.125	.773		.758	1.111	.107	.022
12	3/4	1-1/16-12UNJ-3A	.609	.315	.938	.664	1.375	.945		.864	1.361		.029
16	1	1-5/16-12UNJ-3A	.844	.367	1.188	.913	1.625	1.195		.911	1.599	.125	
20	1-1/4	1-5/8-12UNJ-3A	1.078	.378	1.501	1.147	1.875	1.506		.958	1.849		
24	1-1/2	1-7/8-12UNJ-3A	1.312	.451	1.750	1.381	2.125	1.756	.008	1.083	2.095		.044
28	1-3/4	2-1/4-12UNJ-3A	1.547	.441	2.125	1.646	2.500	2.131		1.208	2.468		
32	2	2-1/2-12UNJ-3A	1.781	.395	2.375	1.880	2.750	2.381		1.333	2.718		
40	2-1/2	3-120UNJ-3A	2.281		2.882	2.360	3.250	2.881		1.109	3.218		
48	3	3-1/2-12UNJ-3A	2.781		3.382	2.880	3.750	3.381		1.187	3.718		

(F) ENTIRE STANDARD REVISED

THIS IS A DESIGN STANDARD, NOT TO BE USED AS A PART NUMBER.

Review activities: AIR FORCE — 82 ARMY — WC NAVY — AS
DSA — CS
User activities: ARMY — ME — AT

APPROVED 25 APR 57 REVISED (F) 11 OCT 74

P.A. AIR FORCE- 82	TITLE	MILITARY STANDARD
Other Cust ARMY- WC NAVY- AS	FITTING END, STANDARD DIMENSIONS FOR FLARED TUBE CONNECTION AND GASKET SEAL	MS 33656
PROCUREMENT SPECIFICATION NONE	SUPERSEDES: AND10056	SHEET 2 OF 2

DD FORM 672-1 (Coordinated)

PREVIOUS EDITIONS OF THIS FORM ARE OBSOLETE.

☆ U. S. GOVERNMENT PRINTING OFFICE: 1977-703-020/4847

Source: U.S. Government Printing Office, 703-020/4847 (1977).

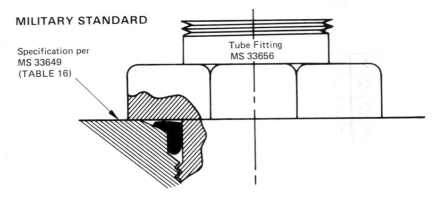

Average Cross-Sectional Squeeze on O-Ring = 38%

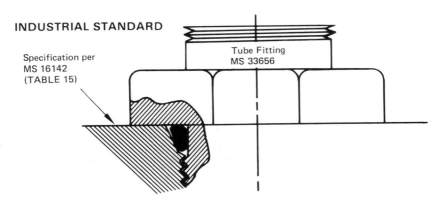

Average Cross-Sectional Squeeze on O-Ring = 26%

Figure 9. Comparison of tube fitting glands

O-ring, especially O-rings of abrasion- and tear-susceptible materials.

The industrial standard is becoming the preferred specification because it subjects the O-ring to less distortion while maintaining a reliable seal. O-rings used in tube fitting glands configured according to the industrial standard (Table 15) have a successful history and have been shown to be reusable many times over. If

the military standard (Table 16) must be used, the O-ring seal
should be replaced prior to each assembly.

VI. FACE-SEAL GLANDS

Table 18 presents dimensions for face-seal glands for applications
up to 1500 psi. The O-ring cross-sectional squeeze ranges from
19 to 32%, while the groove width, G, allows, for a minimum volu-
metric clearance to preclude roll or spiral failure of the O-ring.
The gland depth L, the gland width G, and either the external
or internal diameter of the gland (groove) must be specified. The
external diameter of the gland is specified if the pressure acting
on the O-ring is applied to the inside diameter of the O-ring.
The internal diameter of the gland is specified when the pressure
is applied to the outside diameter of the O-ring. Specifying the
external or internal diameter of the gland, depending on where
the pressure is applied, minimizes movement of the O-ring within
the gland and prevents spiral failure.

The integrity of a face seal depends on an external method of
maintaining compression on the O-ring. Methods include fasteners
such as screws, bolts, or retaining mechanisms such as snap rings
and retaining clips. The designer should not forget to consider
the load required to compress the O-ring when choosing the number of fasteners or type of retaining mechanism. Figures 10 to 14
present a combination of empirical and theoretical data on compression loads for the various O-ring cross sections. The figures
give the force per linear inch of seal circumference as a function
of Shore hardness and percent squeeze. For example, in Fig. 11,
a 0.103-in.-cross-section O-ring of 70° Shore hardness experienc-
ing 26% squeeze will produce a compression load of approximately
15 lb per linear inch of seal circumference. Now, if the O-ring
has a mean diameter of 3.00 in. (size −149 O-ring), the total
force required to maintain the 26% squeeze is

$$F_T = 15 \text{ lb/in.} \times \pi(3.00 \text{ in.}) = 141 \text{ lb}$$

or, using the equation form,

$$F_T = 1.7344e^{8.4720(0.26)}\pi(3.00) = 148 \text{ lb}$$

This force may or may not be significant, depending on the
application of the face seal and the size of the fasteners used.
Forces do become significant when large-diameter O-rings of 90°
Shore hardness material are involved in large squeeze applications.

Table 18. Face-Seal Glands

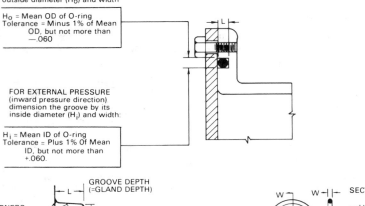

FOR INTERNAL PRESSURE
(outward pressure direction)
dimension the groove by its
outside diameter (H$_O$) and width

H$_O$ = Mean OD of O-ring
Tolerance = Minus 1% of Mean
OD, but not more than
—.060

FOR EXTERNAL PRESSURE
(inward pressure direction)
dimension the groove by its
inside diameter (H$_i$) and width:

H$_i$ = Mean ID of O-ring
Tolerance = Plus 1% Of Mean
ID, but not more than
+.060.

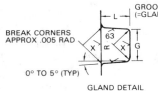

BREAK CORNERS
APPROX .005 RAD

GROOVE DEPTH
(=GLAND DEPTH)

63

0° TO 5° (TYP)

GLAND DETAIL

SURFACE FINISH X:
32 FOR LIQUIDS
16 FOR VACUUM
AND GASES

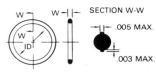

SECTION W-W

.005 MAX.

.003 MAX.

O-ring Size AS 568 Dash No.	W Cross Section		L Gland Depth	Squeeze		G Groove Width		R Groove Radius
	Nominal	Actual		Actual	%	Liquids	Vacuum and Gases	
004 through 050	1/16	.070 ±.003	.050 to .054	.013 to .023	19 to 32	.101 to .107	.083 to .088	.005 to .015
102 through 178	3/32	.103 ±.003	.074 to .080	.020 to .032	20 to 30	.136 to .142	.118 to .123	.005 to .015
201 through 284	1/8	.139 ±.004	.101 to .107	.028 to .042	20 to 30	.177 to .187	.157 to .163	.010 to .025
309 through 395	3/16	.210 ±.005	.152 to .162	.043 to .063	21 to 30	.270 to .290	.236 to .241	.020 to .035
425 through 475	1/4	.275 ±.006	.201 to .211	.058 to .080	21 to 29	.342 to .362	.305 to .310	.020 to .035
Special	3/8	.375 ±.007	.276 to .286	.082 to .108	22 to 28	.475 to .485	.419 to .424	.030 to .045
Special	1/2	.500 ±.008	.370 to .380	.112 to .138	22 to 27	.638 to .645	.560 to .565	.030 to .045

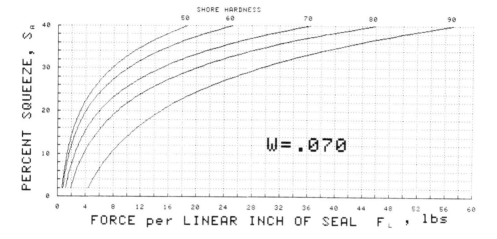

$$F_L = Ke^{CS_A}$$

Shore Hardness	K	C
50	0.5835	8.6712
60	0.6800	9.0298
70	1.0137	8.9588
80	1.6355	8.3366
90	3.8816	6.7245

Figure 10.   O-ring compression loads for 0.070-in. cross section

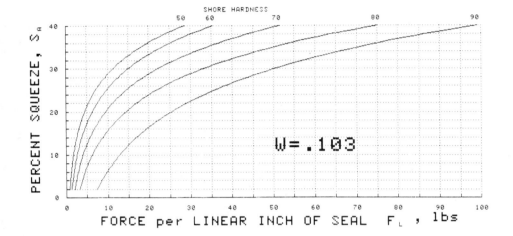

$$F_L = Ke^{CS_A}$$

Shore Hardness	K	C
50	0.6985	9.2670
60	1.1118	8.6513
70	1.7344	8.4720
80	2.7280	8.2919
90	6.4609	6.8259

Figure 11. O-ring compression loads for 0.103-in. cross section

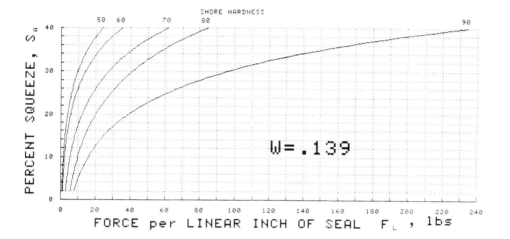

$$F_L = Ke^{CSA}$$

Shore Hardness	K	C
50	0.6450	9.0658
60	0.9034	9.1890
70	2.4197	8.1145
80	4.7819	7.1872
90	6.5035	8.9641

Figure 12. O-ring compression loads for 0.139-in. cross section

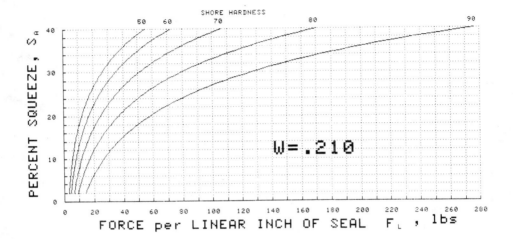

$$F_L = Ke^{CS_A}$$

Shore Hardness	K	C
50	2.6697	7.5627
60	3.8873	7.3138
70	5.6705	7.3260
80	7.9310	7.6663
90	12.1540	7.8110

Figure 13. O-ring compression loads for 0.210-in. cross section

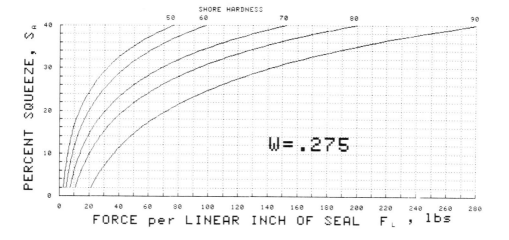

$$F_L = Ke^{CS_A}$$

Shore Hardness	K	C
50	2.3923	8.6554
60	3.9660	8.0304
70	6.2103	8.0001
80	9.3800	7.6454
90	18.5000	6.8000

Figure 14. O-ring compression loads for 0.275-in. cross section

5

Reciprocating Seals—Pistons and Cylinders

I. DESIGN PARAMETERS

Extreme differential pressure and corrosion are the most critical
aspects affecting the design parameters of reciprocating seals.
Extreme pressure requires minimum clearance between the O-ring
housing and piston shaft to ensure against O-ring extrusion.
Side loads on a piston or rod can cause the clearance in the O-
ring gland to be on one side only. If adequate O-ring squeeze
has not been provided, leakage will result, and if excessive clearance is created, extrusion of the O-ring may result. High side
loading on a piston will cause uneven friction on the seal, and if
high enough, the rod or barrel will be galled or scored. Shock
pressures, such as created by sudden stopping of a hydraulic
damping cylinder, are many times greater than the actuation pressures required for normal use and must be considered as critical
design parameters. In many applications, a mechanical lock or
brake should be provided to reduce and/or take up shock loads,
especially after the final piston position (relative to the cylinder)
has been attained.

Corrosion and related contamination by sand, dirt, and mois-
ture can be very detrimental to the sealing efficiency of O-rings
in underwater hydraulic cylinder applications. Equipment having
rods exposed to this hostile environment during operating cycles
should be fitted with scraper and/or wiper rings which prevent
dirt and corrosive products from reaching the O-ring seal and
seal housing. To reduce galvanic corrosion, the usual type of
bearing materials (babbit, bronze, etc.) that most often are dissimilar to the major types of structural materials (steel, stainless
steel, aluminum, etc.), can be replaced by inert materials, such
as nylon and Teflon. Designers using such polymeric materials

must consider and allow for relatively high coefficients of thermal expansion and, in the case of Teflon, cold flow.

In most piston/cylinder applications, the O-ring groove is machined into the male element (piston), since it is usually an easier process and the complications involved in design of a groove in the female part [peripheral compression, as discussed under the theory of rotary seals (Chap. 6)] are eliminated. There are applications, however, where it is functionally more efficient to put the groove in the cylindrical part instead of the piston. If the frictional force of the moving metal surface across the O-ring is in the same direction as the direction of differential pressure, the O-ring will tend to be dragged onto the gap more easily and thus extrude at a much lower (30 to 40%) than normal pressure [3, pp. 6-9]. By placing the groove in the opposite metal part, the friction will work against pressure and reduce the susceptibility to "spiral failure."

See Tables 12 and 14 for design data for military and industrial reciprocating glands.

II. FAILURE AND CAUSES

Spiral failure is a unique type that sometimes occurs on reciprocating O-rings. This type of failure appears as a spiral or corkscrew cut halfway through the O-ring cross section. This seal has the appearance of being twisted while being cut with a knife.

A properly used O-ring slides during all but a small fraction of any reciprocating stroke and does not normally tend to twist or roll. According to general theory, a properly used O-ring does not tend to roll or twist because (1) the differential pressure across the O-ring produces a holding force within the groove due to friction on a larger area, greater than the pulling force produced by the sliding surface (rod or cylinder wall) opposite the groove; (2) the surface finish of the sliding surface is made smoother than that of the groove in order to reduce friction at the sliding surface; (3) running friction between moving parts is always lower than the breakout friction between nonmoving parts; and (4) the torsional resistance of the O-ring tends to resist twisting [2, p. A6-9].

Elastomeric O-rings are not recommended for applications where the cylinder rod has a stroke of more than 12 in. of unsupported length unless extra precautions are taken. Usually, the longer the stroke of a cylinder rod, the greater the eccentricity, bending, side load, and in general, the tendency to produce factors that contribute to wear and/or spiral failure. To

minimize the consequences of eccentricity, bending, and side
loads of a long-stroke cylinder rod, floating glands are used
most often. The object of a floating gland is to allow the piston
or rod bearing (containing the O-ring groove) to pivot, adjust,
or float a small amount, offsetting misalignment. The seal housing
used in the rotary O-ring problem (Fig. 22) is actually a floating
gland or floating seal.

Investigations made by the Parker Seal Company have dis-
closed that spiral failure occurs very often when reciprocating
speeds are less than 1 fpm. The apparent reason for failure at
slow speeds is that the sliding or running seal friction created is
very high and comparable to breakout friction. This high friction
tends to excessively twist the O-ring, especially on low or balanced pressure components, and thus spiral failure occurs. O-
ring seals are not recommended, therefore, for speeds less than
1 fpm when the differential pressure is less than 400 psi [2, p.A6-
10].

The two main design factors that contribute to spiral failure of
reciprocating seals are lack of lubrication and groove shape.
Lack of lubrication between the O-ring and the sliding rod tends
to increase the relative friction and may result in excessive twisting of the O-ring and eventually, spiral failure. If a V-shaped
groove is used, the hydraulic holding force is reduced because the
area on the side of the V-groove is less than at the bottom and
sides of a square groove. Therefore, we can see that when an
unlubricated rod or surface is actuated through a seal contained
in a groove of reduced area, twisting of the seal and spiral failure
has a greater probability of occurring.

III. APPLIED SIDE LOADS AND O-RING RESTORING FORCES

Figure 15 shows a long-stroke piston eccentrically located within a
horizontal cylinder. The weight of the piston acts to increase the
compression of the O-ring at the bottom of the cylinder and decrease the compression on the O-ring at the top of the cylinder.
The piston and cylinder should be dimensioned such that there is
always squeeze on the O-ring at the top of the cylinder, even if
the piston is resting on the bottom of the cylinder wall. If the
piston does not rest on the bottom of the cylinder wall, the O-
ring will have provided enough restoring force to support the
weight of the piston. In such a situation, the designer may need
to determine the effective restoring force as a function of O-ring
size and hardness, the initial squeeze and the eccentricity between

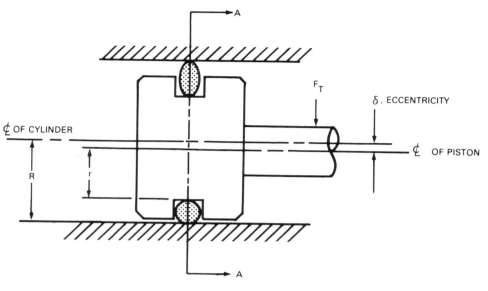

Figure 15.   Side-loaded piston

the piston and the cylinder. Determination of the total restoring force provided by the squeeze on the O-rings in piston-cylinder systems becomes very important when combating externally applied side loads (see Design Examples 4 and 5).

Design Example 4   Maximum Allowable Side Load on Piston

Given:

The piston-cylinder assembly as shown with a 2-329, 70° Shore hardness O-ring.

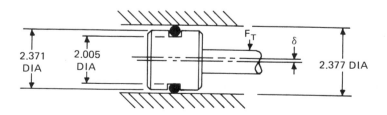

Find:

The maximum allowable side load that can be applied to the piston before the piston encounters the cylinder wall.

Solution:

1. The nominal squeeze on the O-ring: a 2-329 O-ring has a cross-sectional diameter of W = 0.210 in. (see Table 14).

$$S_N = \frac{0.210 - \frac{1}{2}(2.377 - 2.005)}{0.210} = 11.4\%$$

2. The maximum allowable eccentricity before the piston encounters the cylinder wall:

$$\delta_{max} = \frac{1}{2}(2.377 - 2.371) = 0.003 \text{ in.}$$

3. Referring to Appendix 5A, the graph for W = 0.210 in., 70° Shore hardness and $S_N = 10\%$, for a diameter of 2.377 in. and estimating that $\delta = 0.003$ in., we find a total restoring force of approximately

$$F_T = 5 \text{ lb}$$

This is, therefore, the maximum allowable side load.

Design Example 5   Piston Eccentricity

Given:

A cylinder with bore = 8.000 in. diam.
A piston with OD = 7.980 in. diam.
A piston groove diam. = 7.535 in. diam.
An O-ring with W = 0.275 in. and Shore hardness of 60°
A 39-lb side load externally applied to the piston

Find:

The eccentricity between the piston and the cylinder caused by the 39-lb side load. Also find the minimum clearance between the piston and the cylinder.

Solution:

1. The nominal squeeze on the O-ring:

$$S_N = \frac{0.275 - \frac{1}{2}(8.000 - 7.535)}{0.275} = 15.5\%$$

2. Referring to Appendix 5A, the graphs for W = 0.275 in., 60° Shore hardness, and $S_N = 15\%$, for a total force of 39 lb and a diameter of 8.0 in., the eccentricity $\delta = 0.008$ in.

3. The minimum clearance between the piston and the cylinder:

$C = \frac{1}{2}(8.000 - 7.980) = 0.010$ in.

Minimum clearance $= 0.010 - 0.008 = 0.002$ in.

In order to determine the restoring force provided by the O-ring, a relationship between the squeeze S_A at any point around the O-ring and the amount of eccentricity δ between the piston and cylinder must first be derived. Figure 16 is a cross-sectional view of the O-ring of Fig. 15 with r the minimum diameter of the

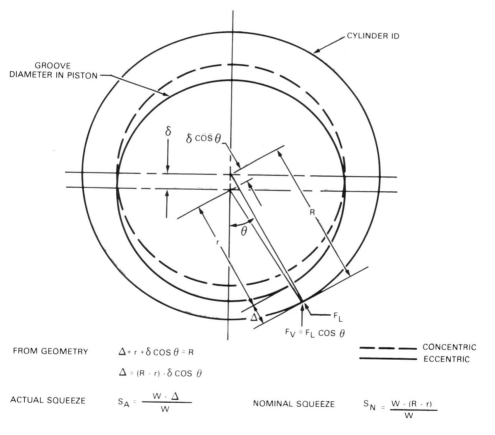

FROM GEOMETRY $\Delta + r + \delta \cos \theta = R$

$\Delta = (R - r) - \delta \cos \theta$

ACTUAL SQUEEZE $S_A = \dfrac{W - \Delta}{W}$ NOMINAL SQUEEZE $S_N = \dfrac{W - (R - r)}{W}$

SECTION AA

Figure 16. Piston gland cross section

O-ring groove in the piston, R the radius of the cylinder, and Δ the gland depth incurred by the O-ring at a specific location θ degrees. The actual squeeze incurred by the O-ring (percentage squeeze divided by 100%) is therefore defined by

$$S_A = \frac{W - \Delta}{W} \tag{1}$$

where W is the cross-sectional diameter of the O-ring. From the geometry shown in Fig. 16.

$$\Delta + r + \delta \cos \theta = R \tag{2}$$

and then

$$\Delta = (r - r) - \delta \cos \theta \tag{3}$$

Substituting into equation (1), we get

$$S_A = \frac{W - R + r + \delta \cos \theta}{W} \tag{4}$$

Because the nominal squeeze on the O-ring can be defined by

$$S_N = \frac{W - (R - r)}{W} \tag{5}$$

equation (4) can be redefined by

$$S_A = S_N + \frac{\delta}{W} \cos \theta \tag{6}$$

From the empirical and theoretical data presented in Figs. 10 to 14, the general equation for the force F_L required to squeeze an O-ring seal is given by

$$F_L = Ke^{C(S_A)} \tag{7}$$

where K and C are constants based on the O-ring cross section W and the Shore hardness. This exponential equation gives the force per unit length of seal circumference.

Substituting equation (6) into equation (7) gives

$$F_L = Ke^{C[S_N + (\delta/W)\cos\theta]} \tag{8}$$

From Fig. 16, the vertical component of the force F_L as a function of θ is

$$F_V = F_L \cos \theta \tag{9}$$

and then

$$F_V = Ke^{C[S_N + (\delta/W)\cos\theta]}\cos\theta \qquad (10)$$

This represents the vertical restoring force at any point on the O-ring tending to support the piston in the cylinder. The summation of all the vertical restoring forces around the O-ring represents the total restoring force tending to make the piston concentric within the cylinder. This is usually accomplished by integration. However, since equation (10) is not directly integrable using conventional calculus, numerical integration was necessary for a solution. The numerical integration of equation (10) required the summation of all the vertical components around the circular O-ring in 1-in. increments because F_L in equation (7) depends on the unit length being 1-in. increments. (To increase the accuracy of the summation, the 1-in. increment was actually decreased to 1/10 in., with the total summation being similarly compensated for this smaller interval.) A computer was incorporated to calculate the total concentric restoring force F_T by summing the increments n around the O-ring. For n equal to 1-in. increments, and because 2π radians equals the circumference of the O-ring, the incremental angle θ in radians is

$$\theta = \frac{2n}{D} \text{ where D is the diameter of the O-ring}$$

Therefore, the total concentric restoring force is

$$F_T = \sum_{i=0}^{n} Ke^{C[S_N + (\delta/W)\cos(2n/D)]}\cos\frac{2n}{D} \qquad (11)$$

The total concentric restoring force can be determined numerically for a particular piston-cylinder application, given the O-ring diameter, its Shore hardness, the cross-sectional squeeze on the O-ring when the piston is concentric within the cylinder, and the actual eccentricity between the piston and cylinder. The total restoring forces for various O-rings up to 18 in. in diameter and up to an eccentricity $\delta = 0.016$ in. are presented in graphs in Appendix 5A. Graphs are given for 10%, 15%, 20%, and 25% nominal cross-sectional squeeze for each of the five common O-ring cross-sectional sizes and for each of the five common Shore hardnesses. Design Examples 4 and 5 describe the use of these graphs.

IV. DYNAMIC FRICTION

The surprising feature of elastomeric O-rings when used in dynamic sealing applications is their ability to wear metal parts. Friction between an O-ring and any bearing surface is caused by the asperities, or microscopic hills and valleys, in the surface interface. Even though an O-ring is much softer than the metal surface it slides against, the O-ring will wear away the asperities of the metal as it "flows" past its irregular surface. The result is usually a polishing effect on the metal surface, but if the metal surface is soft, excess wear to the point of leakage may occur.

To minimize wear and system hysteresis, the designer wants to reduce dynamic friction between O-rings and the surface of the moving part. Design features that accomplish this are smooth finishes below 16 rms on moving parts, increasing speed of moving parts, decreasing cross section of O-ring, decreasing O-ring cross-sectional squeeze, decreasing hardness of O-ring, increasing environmental temperature in piston applications, decreasing environment temperature in rod applications, and lubricating the moving part. Each feature is easily accomplished, but some have penalties, especially in reciprocating applications: the smoother finishes are more costly; decreasing the cross section of the O-ring, decreasing the hardness of the O-ring, and increasing the environmental temperature in piston applications, and decreasing the temperature in rod applications may result in spiral failure of the O-ring. This type of O-ring failure was discussed earlier. As always, the designer must make compromises that ensure proper O-ring function without excess friction.

A. Nomogram Method

The amount of dynamic friction caused by an elastomeric O-ring seal used in a reciprocating application can be approximated using nomograms (Figs. 17, 18). These were devised from theoretical modeling and empirical data. The empirical data were obtained from tests using standard-size elastomeric O-rings (other than Teflon) reciprocating against 15-rms finished chrome-plated surfaces at speeds greater than 1 fpm and lubricated with hydraulic oil MIL-H-5606, at environmental room temperatures [2, p.A6-6]. Figure 17 gives the O-ring friction caused by the differential pressure across the O-ring sealing surface, while Fig. 18 gives the O-ring friction caused by the squeeze incurred by the O-ring seal. The total dynamic friction is the sum of the dynamic frictions from both nomograms. The static friction of the O-ring seal is usually

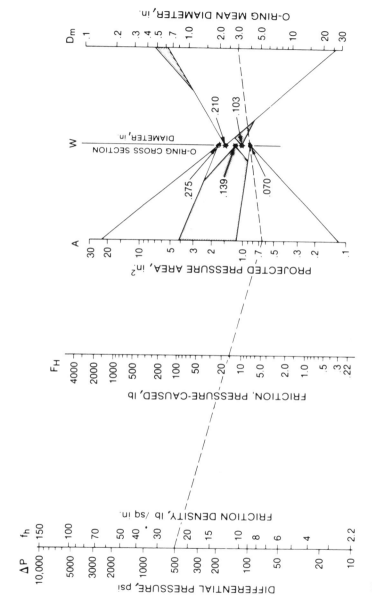

Figure 17. Nomogram 1: O-ring friction due to differential pressure

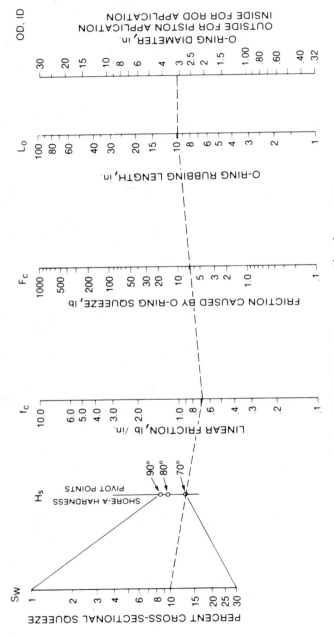

Figure 18.   Nomogram 2:   O-ring friction due to cross-sectional squeeze

between 1.5 and 3 times the total dynamic friction. The dynamic hysteresis or system hysteresis is defined as the dynamic force spent in friction divided by the actual force acting on the piston. Design Example 6 shows how to use the nomograms.

In Design Example 6 the tables recommend an O-ring of 0.210-in. cross-sectional diameter for a nominal 3-¼-in. piston-cylinder diameter. This results in a nominal dynamic friction force of 52 lb. Notice that the O-ring stretch is only 1.05% and has little affect (1.0%) on the actual O-ring cross-sectional diameter.

The dynamic friction force of 52 lb can be reduced by designing an equivalent system using a 0.070-in.-diameter O-ring cross section instead of the 0.210-in.-diameter O-ring cross section. Such a system is presented in Design Example 7. The dynamic friction of this system is only 23 lb, half that calculated for the 0.210-in.-diameter O-ring. The piston groove is dimensioned to produce a nominal 5% O-ring stretch, compared to 1.05% stretch for the 0.210-in.-diameter O-ring. This is accomplished so that the O-ring will hug the piston with greater force to provide greater resistance to spiral failure. Increasing the O-ring stretch to 5% is recommended when designing for an O-ring cross section smaller than that normally specified in the industrial tables.

The original O-ring cross section, 0.070 in., is used in determining the percent cross-sectional squeeze to be approximately 10%. The actual squeeze on the installed O-ring cross section is later checked to ensure proper O-ring squeeze. This piston-cylinder design results in a minimum squeeze of only 1.54%, while the maximum squeeze is only 11.0%. This design therefore represents a minimum-friction system. Design Example 8 solves the same problem using the actual worst-case O-ring squeeze of 11.0% in the equation method for comparison with the nomogram method. The difference is usually within normal engineering tolerance.

Design Example 6   Friction of the Reciprocating Elastomeric Seal 2-336 O-Ring

Problem statement:

Select an O-ring for and dimension the pertinent features of a nominal 3-¼-in.-diameter piston-cylinder system. Then determine the nominal dynamic and static frictions and the nominal dynamic hysteresis of the reciprocating system for a differential pressure of 500 psi.

Solution:

1. From the Parker Seal Company *O-Ring Handbook for Industrial Specification for Reciprocating O-Ring Seals* (Table 14), the following dimensions are recommended:

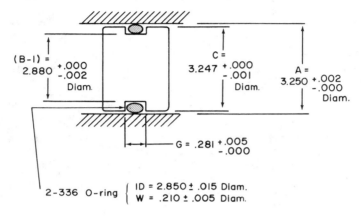

$(B-1) =$ 2.880 $^{+.000}_{-.002}$ Diam.

$C =$ 3.247 $^{+.000}_{-.001}$ Diam.

$A =$ 3.250 $^{+.002}_{-.000}$ Diam.

$G = .281 ^{+.005}_{-.000}$

2-336 O-ring $\begin{cases} ID = 2.850 \pm .015 \text{ Diam.} \\ W = .210 \pm .005 \text{ Diam.} \end{cases}$

2. Nominal dynamic friction:
   a. Using Fig. 17, where

   $D_m = 2.850 + 0.210 = 3.060$ in.

   $W = 0.210$ in.

   $\Delta P = 500$ psi

   we find that $A = 2$ in.^{2}, $f_h = 24$ lb/in.^{2}, and $F_H = 45$ lb.
   b. Using Fig. 18, where

   O-ring OD $= 2.850 + 2(0.210) = 3.270$

   the percent cross-sectional squeeze is

   $$S_W = \frac{0.210 - (3.250 - 2.880)/2}{0.210} \times 100\% = 11.9\%$$

   and the O-ring Shore hardness is $H_S = 70°$, we find that $L_0 = 10$ in., $f_c = 0.72$ lb/in., and $F_C = 7.0$ lb.
   c. Then the total nominal dynamic friction is

   $F = 45 + 7.0 = 52$ lb (Answer)

3. The nominal static friction may be approximated as

   $3 \times 52 = 156$ lb max. (Answer)

4. The nominal dynamic hysteresis of the system is

$$\frac{\text{Nominal dynamic friction}}{\text{Nominal force acting on piston}} \times 100\%$$

$$= \frac{52 \text{ lb}}{(500 \text{ lb/in.}^2)(\pi/4)(3.250/\text{in.})^2} \times 100\%$$

$$= 1.25\% \text{ (Answer)}$$

Check on nominal squeeze:

1. Nominal percentage O-ring diametral stretch:

$$\frac{2.880 - 2.850}{2.850} \times 100\% = 1.05\%$$

2. Actual O-ring cross-sectional diameter: According to Fig. 6, for 1.05% diametral stretch, the percentage reduction in cross-sectional diameter is 1.0%, and therefore the actual O-ring cross-sectional diameter is

$$W' = 0.210 \times (1 - 0.01) = 0.208 \text{ in.}$$

3. Nominal O-ring squeeze:

$$\frac{0.208 - (3.250 - 2.880)/2}{0.208} \times 100\% = 11.1\%$$

Design Example 7 Friction of Reciprocating Elastomeric Seal

Given:

The hydraulic cylinder with an AS568-041 O-ring and a maximum working pressure of 500 psi across the piston:

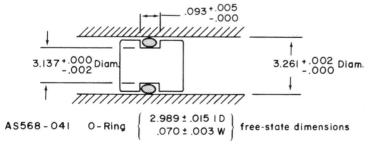

AS568-041   O-Ring   { 2.989 ± .015 ID   .070 ± .003 W }   free-state dimensions

Find:

The dynamic and static friction of the reciprocating piston and the dynamic hysteresis of the system.

Solution:

1. Dynamic friction:
   a. Using Fig. 17, where

   $$D_m = 2.989 + 0.070 \cong 3.0 \text{ in.}$$

   $$W = 0.070 \text{ in.}$$

   $$\Delta P = 500 \text{ psi}$$

   we find $A = 0.65$ in.^{2}, $f_h = 24$ lb/in.^{2}, and $F_H = 16$ lb.
   b. Using Fig. 18, where

   $$\text{O-ring OD} = 2.989 + 2(0.070) = 3.129 \text{ in.}$$

   the percent cross-sectional squeeze is

   $$S_W = \frac{0.070 - (3.261 - 3.137)/2}{0.070} \times 100\% \cong 10\%$$

   and O-ring Shore hardness $H_S = 70°$, we find that $L_0 = 9.8$ in., $f_c = 0.68$ lb/in., and $F_c = 6.7$ lb.
   c. Then the total dynamic friction is

   $$F = 16 + 6.7 \cong 23 \text{ lb}$$

2. The static friction may be approximated as

   $$3 \times 23 = 69 \text{ lb max.}$$

3. The dynamic hysteresis of the system is

   $$\frac{\text{Dynamic friction}}{\text{Force acting on the piston}} \times 100\%$$

   $$= \frac{23 \text{ lb}}{(500 \text{ lb/in.}^2)(\pi/4)(3.261 \text{ in.})^2} \times 100\% = 0.55\%$$

Check on actual squeeze:

1. Percentage O-ring diametral stretch:
   a. Maximum piston groove, minimum O-ring ID:

   $$\frac{3.137 - 2.974}{2.974} \times 100\% = 5.5\% \text{ max.}$$

   b. Minimum piston groove, maximum O-ring ID:

$$\frac{3.135 - 3.004}{3.004} \times 100\% = 4.4\% \text{ min.}$$

2. O-ring cross section as stretched on piston: From the plot of percent reduction in cross-section diameter (Fig. 6), we find for 5.5% diametral stretch a maximum O-ring cross-section reduction of 4%; therefore,

$$W' = (0.070 \pm 0.003)(1 - 0.04) = 0.0672 \pm 0.0029$$

$$= 0.0643 \text{ in. min., } 0.0701 \text{ in. max.}$$

From the same plot, we find for 4.4% diametral stretch a maximum O-ring cross-section reduction of 3%; therefore,

$$W' = (0.070 \pm 0.003)(1 - 0.03) = 0.0679 \pm 0.0029$$

$$= 0.0650 \text{ in. min., } 0.0708 \text{ in. max.}$$

3. Percent cross-sectional squeeze when O-ring is installed:

Maximum squeeze; large O-ring, small cylinder bore, small piston groove:

$$\frac{0.0708 - (3.261 - 3.135)/2}{0.0708} \times 100\% = 11.0\%$$

Minimum squeeze; small O-ring, large cylinder bore, small piston groove:

$$\frac{0.0650 - (3.263 - 3.135)/2}{0.0650} \times 100\% = 1.54\%$$

When designing a minimum friction system, one should always check that the minimum O-ring squeeze is sufficient to maintain the seal.

B. Equation Method

The dynamic friction caused by an elastomeric O-ring seal used in a reciprocating application can be calculated using the following equations:

1. $F = F_H + F_C$, total dynamic friction
2. $F_H = Af_h$, friction due to differential pressure across O-ring cross section; $A = \pi D_m W$, projected pressure area of the O-ring; D_m = mean diameter and W = original O-ring cross-sectional diameter; $f_h = 0.545 \, (\Delta P)^{0.61}$, friction density; ΔP = differential pressure across O-ring cross section.

3. $F_C = L_0 f_c$, friction due to O-ring cross-sectional squeeze; $L_0 = \pi(\text{OD or ID})$, O-ring rubbing length; OD = outside diameter for piston application, ID = inside diameter for rod application; $f_c = (-0.884 + 0.0206 H_S - 0.0001 H_S{}^2) S_W$, linear friction, H_S = Shore hardness of O-ring, S_W = actual squeeze of O-ring cross section in percent.

4. Therefore, $F = \pi D_m W[0.545(\Delta P)^{0.61}] + \pi(\text{OD or ID})(-0.884 + 0.0206 H_S - 0.0001 H_S{}^2) S_W$.

The equation in 4 determines the total dynamic friction of an O-ring seal reciprocating against a 15-rms finished metal surface at speeds greater than 1 fpm and lubricated with hydraulic oil MIL-H-5606 at environmental room temperatures. Design Example 8 shows how to use the equation. The equation method 4 may not be considered reliable for differential pressures ΔP greater than 4000 psi, or O-ring cross-sectional squeezes greater than 30%.

Design Example 8 Friction of Reciprocating Elastomeric Seal

For the same size 2-041 O-ring that was used in Design Example 7, we may substitute the following into the equation in 4:

$$
2\text{-}041 \text{ O-ring} \begin{cases} D_m = 0.989 + 0.070 = 3.059 \text{ in.} \\ W = 0.070 \text{ in.} \\ \text{OD} = 2.989 + 2(0.070) = 3.129 \text{ in.} \\ H_S = 70° \end{cases}
$$

$\Delta P = 500$ psi

$S_W = 11.0\%$ actual squeeze from Design Example 7

Reciprocating piston application; light lubricant, 15-rms surface finish, speeds beyond 1 fpm.

$$F = F_H + F_C$$

$$F_H = \pi(3.059)(0.070)[0.545(\Delta P)^{0.61}] = 16.24 \text{ lb}$$

$$F_C = \pi(3.129)[-0.884 + 0.0206(70) - 0.0001(70)^2]11.0 = 7.35 \text{ lb}$$

$$F = 23.6 \text{ lb.}$$

Notice that this answer is only 2.6% different than that determined by using the nomograms. This is mostly because the actual maximum O-ring squeeze of 11.0% was used, whereas in the nomogram example, 10% squeeze was used for reasons of relative comparison. It is always good design practice to use at least 10% squeeze when approximating the least maximum squeeze to be incurred by a reciprocating O-ring, see the following section.

C. Minimum Friction

If one were to design for the absolute minimum friction for an O-ring used in a reciprocating application, the tolerances of the piston groove, cylinder bore, and O-ring itself would result in at best 0% minimum to 9.6% maximum O-ring cross-sectional squeeze. Design Examples 9 and 10 present absolute minimum friction designs for the two size extremes: very smallest 0.070 O-ring, 2-006, and the very largest, 2-475. Both examples show a minimum O-ring squeeze of zero. The very small 2-006 O-ring results in a maximum 17.81% O-ring squeeze. This occurs when a large-tolerance 0.070-in. cross section is installed in a small cylinder bore and a small piston groove. The very large 2-475 O-ring results in a maximum 7.83% O-ring squeeze when a large-tolerance 0.281-in. cross section is installed in a small cylinder bore and a small piston groove.

When approximating O-ring friction of even the best minimum friction O-ring design, it is a good rule of thumb to use no less than 10% O-ring squeeze. This should be applied in nomogram 2 (Fig. 18) and in calculations when computing worst-case dynamic friction for the very best minimum-friction O-ring design. Of course, when the actual percent O-ring squeeze is known, whether greater or less than 10%, it should be used to determine the actual dynamic friction.

V. STATIC FRICTION

The static friction of an elastomeric O-ring system is defined as the force required to initiate motion. Sometimes called "breakout friction" or "holding force," this force is the impetus thrust that breaks the O-ring from a static seal into a dynamic seal. This static friction force is normally between 1.5 and 3 times the force required to maintain motion, the dynamic friction force. The static friction of an O-ring system will increase to a constant maximum as the delay time between stop and start of motion increases. For example, consider a piston containing an O-ring seal moving down a cylinder bore. The piston requires a 3-lb force to maintain its motion, the dynamic friction of the O-ring

seal. Now, assume that the piston is stopped. The piston
stays motionless for 40 min and then is required to move. Because
the rubber O-ring seal has now had time to flow into the microfine
surface irregularities of the cylinder bore, the force required to
start piston motion will be 5 lb. If motion had been delayed for
20 hr, the static friction force would have been 8 lb. The static
friction would increase to an asymptotic limit of 9 lb at about 200 hr
delay time. Any delay time over 200 hr would not increase the
static friction beyond 9 lb, because the rubber would have flowed
into all the available surface grooves and irregularities by this
time, and additional time could not create any addition holding
force by the rubber O-ring.

The static friction of an elastomeric O-ring system depends
on the same factors that influence dynamic friction. Such factors
as the surface finish on the moving part, the squeeze on the
O-ring, the hardness of the O-ring material, the type and amount
of lubrication, and the fluid temperature and pressure affect the
static friction in direct proportion to their effect on the dynamic
friction. A design for minimum static friction would include a surface finish of 8 rms or less on the moving part; the minimum
squeeze on the O-ring without allowing leakage, 1.0%; the softest
available O-ring, 60 durometer; and a lightweight and O-ring-
compatible lubricant. The environmental conditions should be
controlled when possible such that low temperatures are not incurred, differential pressure does not increase, and motion-delay
time (time parts are at rest) is kept to a minimum.

VI. SYSTEM HYSTERESIS

The system hysteresis of an elastomeric O-ring system is defined
as the ratio of the O-ring friction to the motivating force. In the
case of a piston-cylinder system, the system's dynamic hysteresis
would be the dynamic O-ring friction divided by the force acting
on the piston. This system's static hysteresis would be the static
O-ring friction divided by the force acting on the piston, usually
being 1.5 to 3 times the value of the dynamic hysteresis. Most
reciprocating elastomeric seal applications are designed for minimum system hysteresis. Figure 19 shows that system hysteresis
becomes a problem for small O-rings in low-pressure applications.
Figure 20 shows that system hysteresis also becomes a problem
for small O-rings incurring large cross-sectional squeeze. These
figures are presented to give the designer a comparative view as

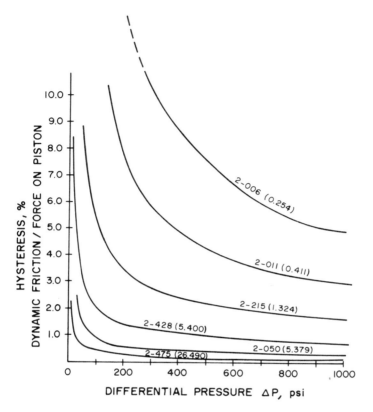

Figure 19. System hysteresis for constant 10% O-ring squeeze S_W

to the effect of these parameters on system hysteresis. The O-ring outside diameter appears in parentheses with the O-ring dash number.

VII. DESIGN FOR ABSOLUTE MINIMUM FRICTION

The following two examples show how to design for the absolute minimum amount of O-ring squeeze and therefore minimum friction for an O-ring used in a reciprocating application. The tolerances used for the O-rings are class 1.

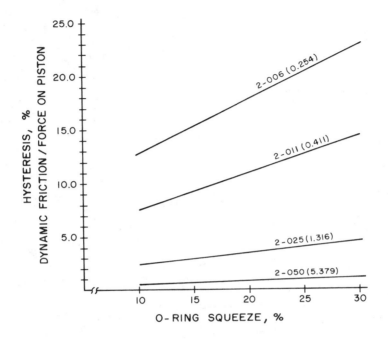

Figure 20. System hysteresis for constant differential pressure ΔP of 200 psi

Design Example 9 Minimum Friction for Very Small O-Ring

For the very small 2-006 O-ring:

$W = 0.070 \pm 0.003$ in

$ID = 0.070 \pm 0.005 \begin{cases} 0.075 \text{ in. max.} \\ 0.065 \text{ in. min.} \end{cases}$

The piston groove diameter must be at least 0.075 in. so that the piston contacts the ID of the largest O-ring. Thus the piston groove should be dimensioned as

(B-1) nomenclature of table $= 0.077 \begin{cases} + 0.000 \text{ in.} \\ - 0.002 \text{ in.} \end{cases} = $ diam (Answer)

To determine the dimensions of the cylinder bore A, we must consider the worst-case limits of the piston groove and the O-ring cross section W and W':

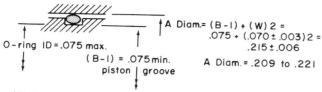

Minimum piston groove and maximum O-ring ID

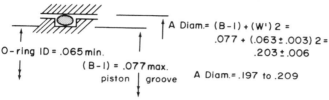

Minimum piston groove and minimum O-ring ID

The reduced O-ring cross-sectional diameter W' was determined by calculating the actual percentage diametral stretch of the ID of the O-ring as

$$\frac{0.077 - 0.065}{0.065} \times 100\% = 18.5\%$$

and finding the corresponding percentage cross-section reduction of 9.7% from Fig. 6, and then calculating:

W' = (0.070 ± 0.003)(1 − 0.097) = 0.063 ± 0.003 in.

(W' = 0.060 to 0.066 in. when the O-ring is stretched on a maximum piston groove.) Thus, since the minimum cylinder bore diameter A equals 0.197 in., the cylinder bore should be dimensioned as

$$A = 0.195 \begin{cases} + 0.002 \text{ in.} \\ - 0.000 \text{ in.} \end{cases} = \text{diam. (Answer)}$$

To check the actual squeeze:

Maximum squeeze: large O-ring installed in small piston groove diameter and minimum cylinder bore:

$$\frac{0.073 - (0.195 - 0.075)/2}{0.073} \times 100\% = 17.81\%$$

Minimum squeeze: small O-ring installed in large piston groove diameter and maximum cylinder bore:

$$\frac{0.060 - (0.197 - 0.077)/2}{0.060} \times 100\% = 0\%$$

Design Example 10 Minimum Friction for Very Large O-Ring

For the very large 2-475 O-ring:

W = 0.275 ± o.006 in.

ID = 25.940 ± 0.060 $\begin{cases} 26.000 \text{ in. max} \\ 25.880 \text{ in. min.} \end{cases}$

The piston groove diameter must be at least 26.000 in. so that the piston contacts the ID of the largest O-ring. Thus the piston groove should be dimensioned as

(B-1) = 26.010 $\begin{cases} + 0.000 \text{ in.} \\ - 0.010 \text{ in.} \end{cases}$ = diam. (Answer)

To determine the dimensions of the cylinder bore A, we must consider the worst-case limits of the piston groove and the O-ring cross sections W and W':

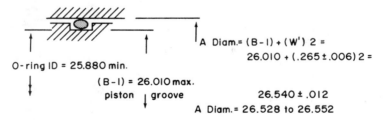

A Diam.= (B-1) + (W) 2 =
26.000 + (.275 ± .006)2 =

O-ring ID = 26.000 max.

(B-1) = 26.000 min.
piston   groove

26.550 ± .012
A Diam.= 26.538 to 26.562

Minimum piston groove and maximum O-ring ID

A Diam.= (B-1) + (W') 2 =
26.010 + (.265±.006)2 =

O-ring ID = 25.880 min.

(B-1) = 26.010 max.
piston   groove

26.540 ± .012
A Diam.= 26.528 to 26.552

Maximum piston groove and minimum O-ring ID

The reduced O-ring cross-sectional diameter W' was determined by calculating the actual percentage diametral stretch of the ID of the O-ring as

$$\frac{26.010 - 25.880}{25.880} \times 100\% = 5\%$$

and finding the corresponding percentage cross-section reduction of 3.5% from Fig. 6, and then calculating:

$$W' = (0.275 \pm 0.006)(1 - 0.035) = 0.265 \pm 0.006 \text{ in.}$$

($W' = 0.259$ to 0.271 in. when an O-ring is stretched on a maximum piston groove.) Thus, since the minimum cylinder bore diameter A equals 26.528 in., the cylinder bore should be dimensioned as

$$A = 26.518 \begin{cases} + 0.010 \text{ in.} \\ - 0.000 \text{ in.} \end{cases} = \text{diam. (Answer)}$$

To check the actual squeeze:

Maximum squeeze: large O-ring installed in small piston groove diameter and minimum cylinder bore:

$$\frac{0.281 - (26.518 - 26.000)/2}{0.281} \times 100\% = 7.83\%$$

Minimum squeeze: small O-ring installed in large piston groove diameter and maximum cylinder bore:

$$\frac{0.259 - (26.528 - 26.010)/2}{0.259} \times 100\% = 0\%$$

From these last two examples, it should be apparent that for the very best minimum-friction O-ring design, the maximum O-ring squeeze can not be less than 7.83%. Therefore, when approximating O-ring friction of even the best minimum-friction O-ring design, it is a good rule of thumb to use no less than 10% O-ring squeeze.

APPENDIX 5A. TOTAL RESTORING FORCES FOR ECCENTRIC PISTONS

(Refer to Section III, p. 110, for use and derivation.)

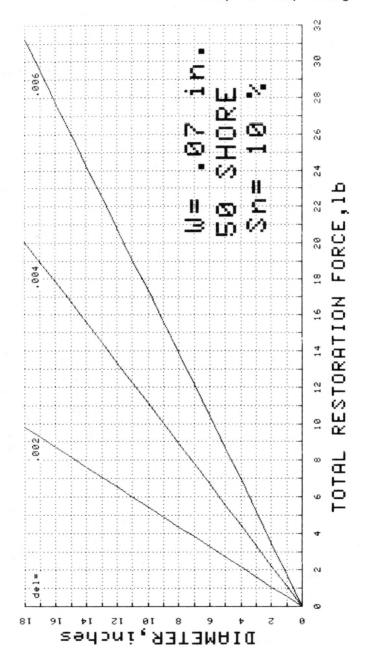

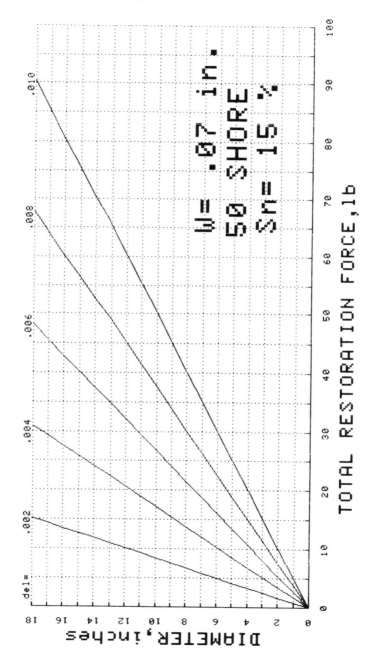

TOTAL RESTORATION FORCE, lb

DIAMETER, inches

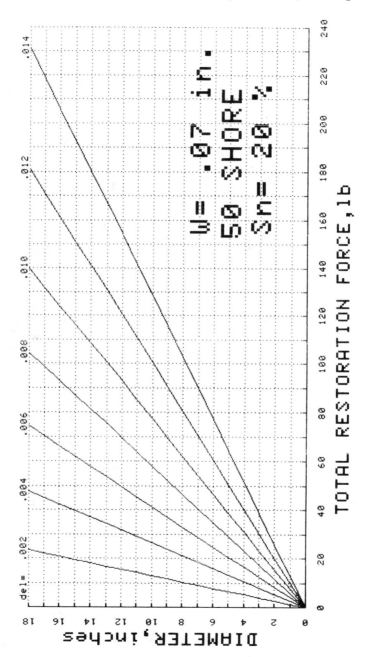

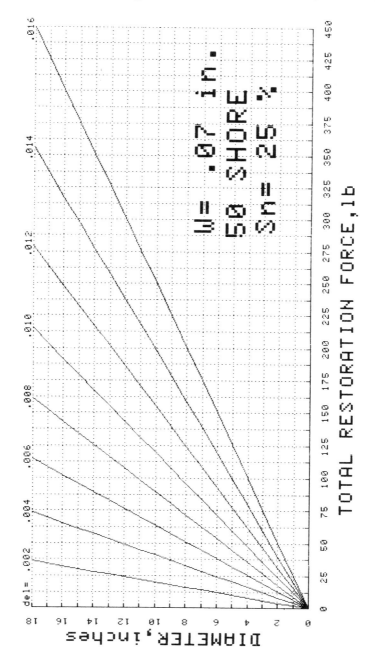

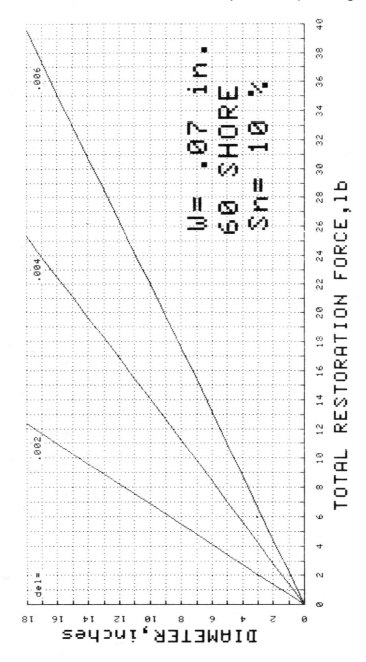

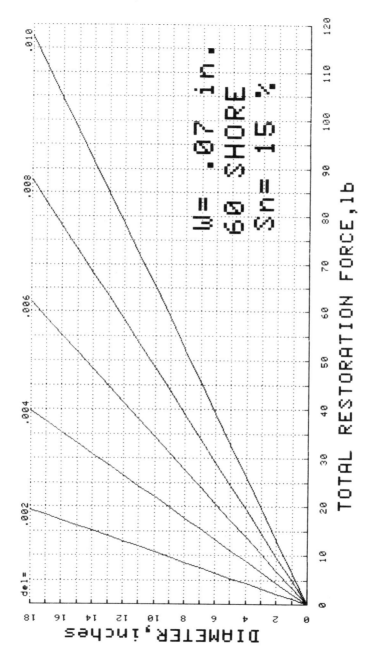

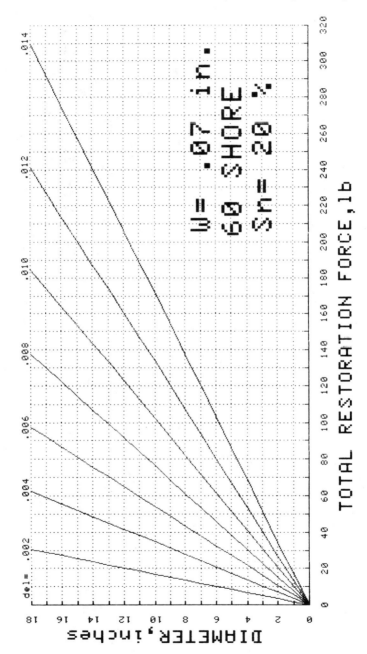

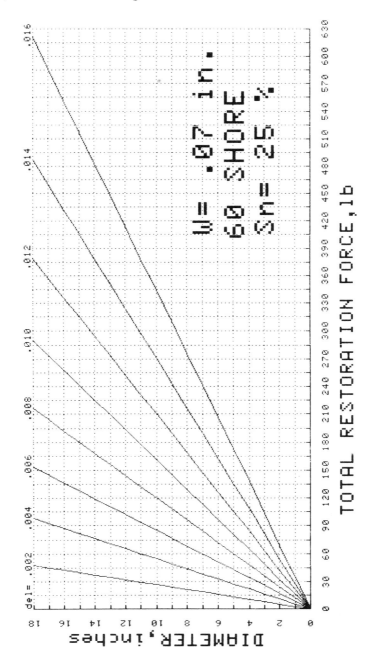

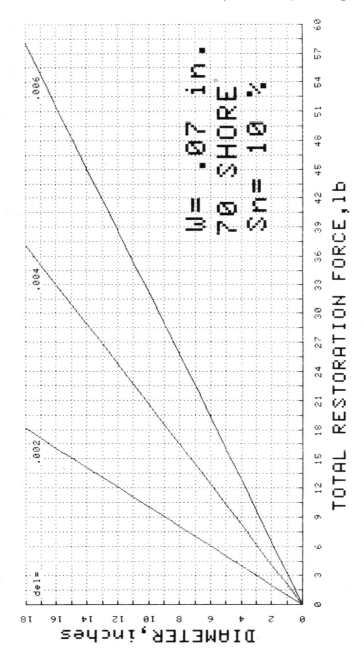

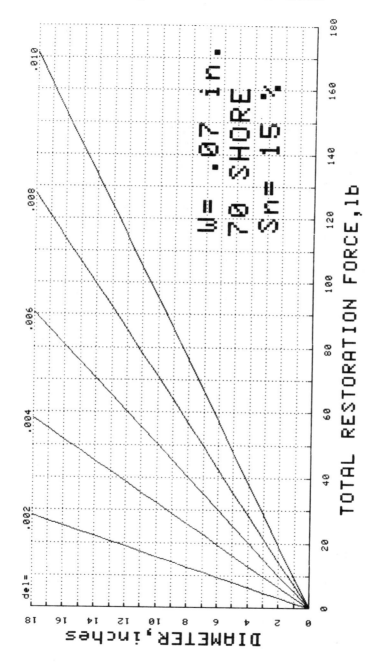

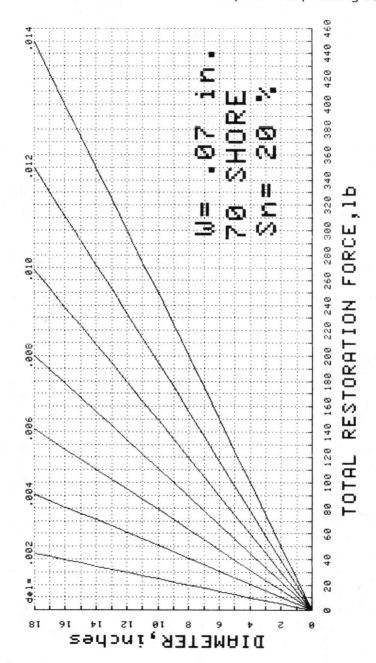

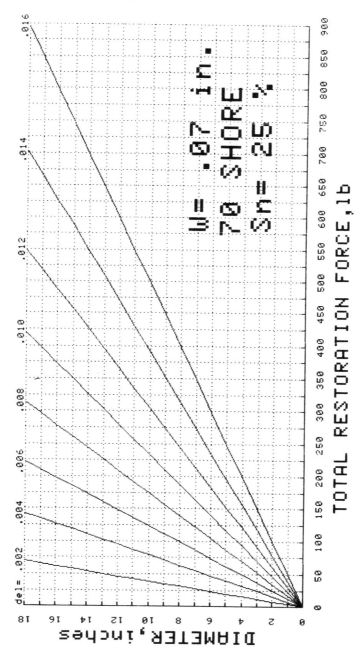

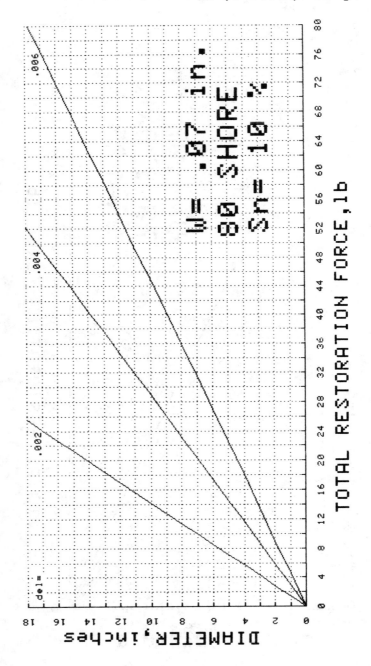

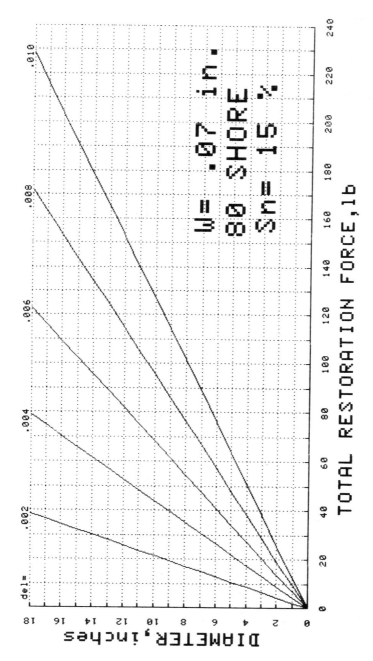

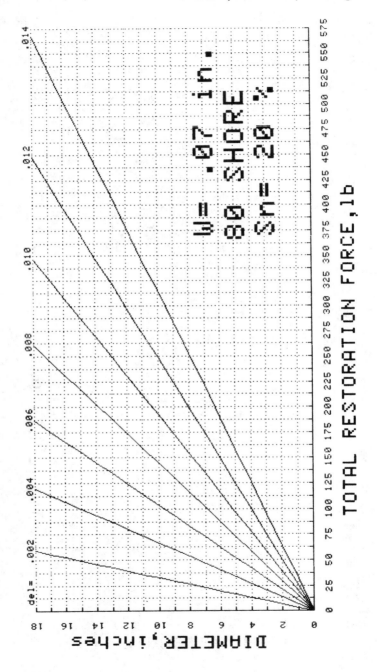

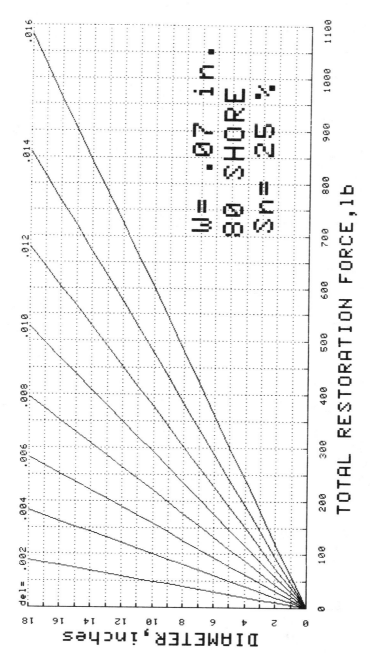

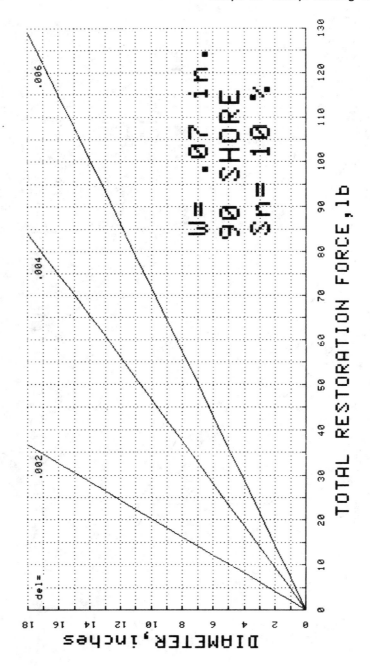

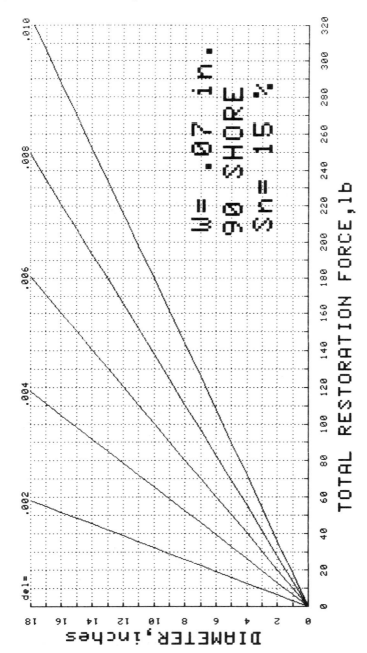

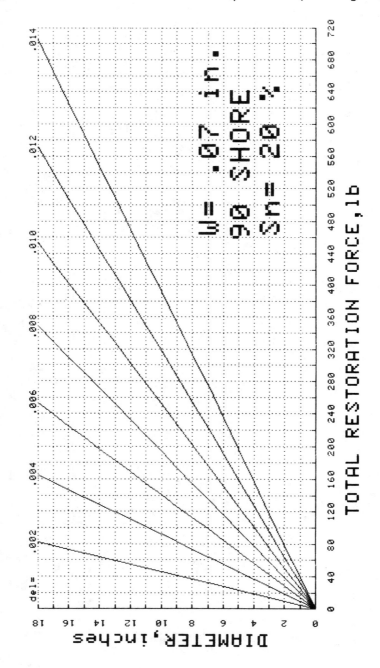

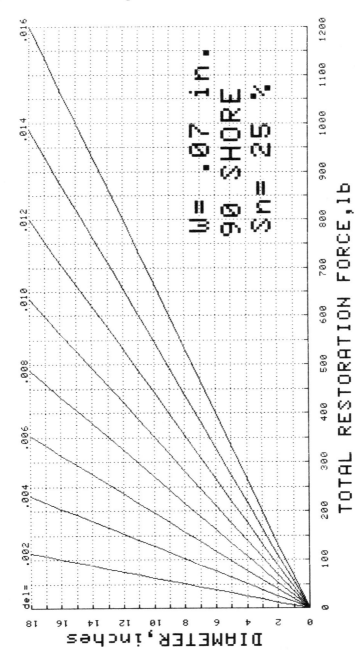

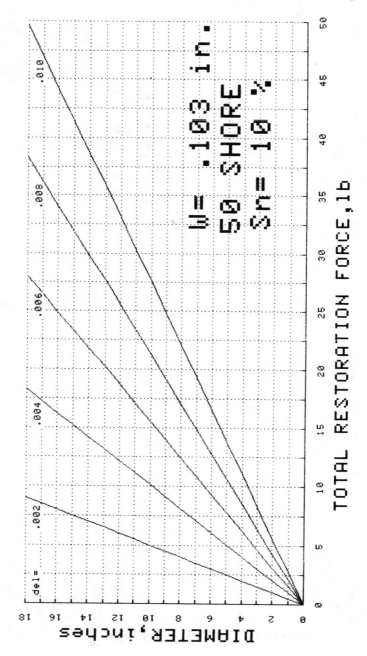

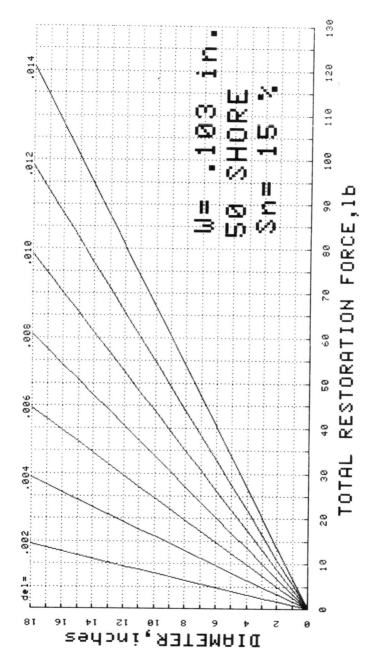

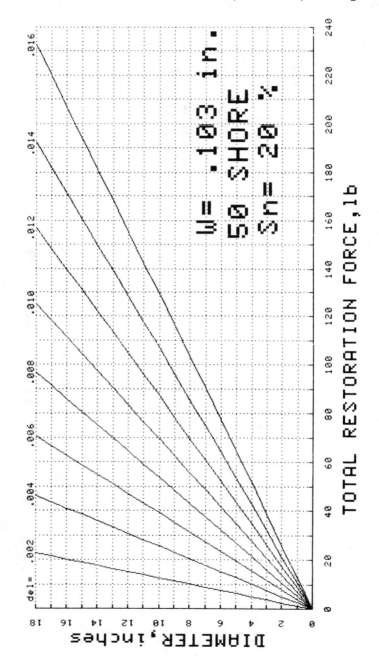

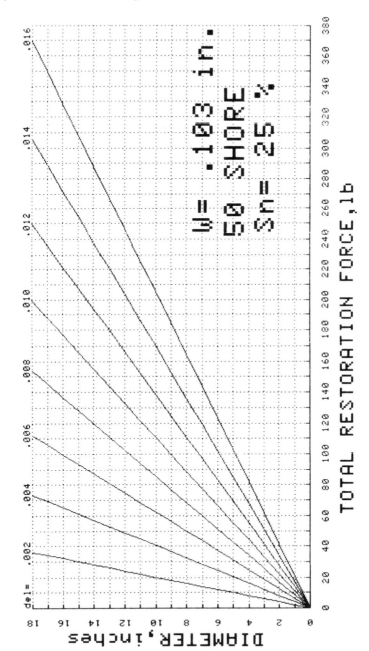

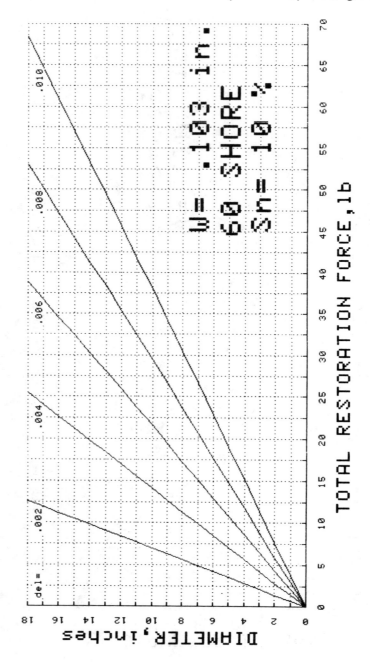

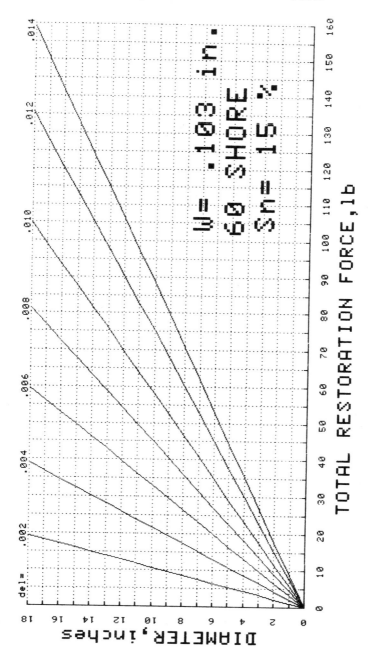

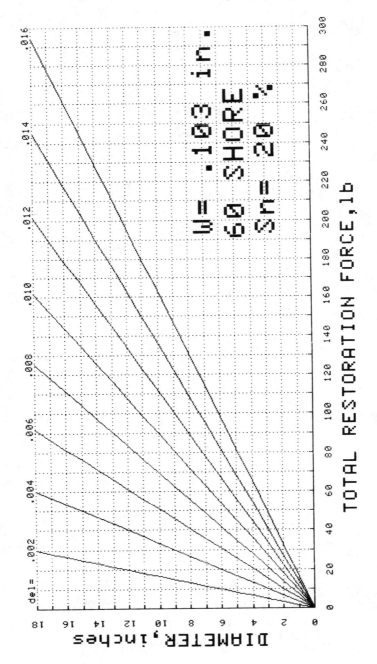

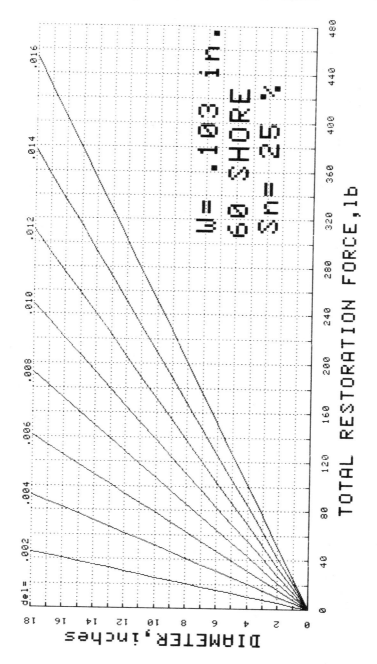

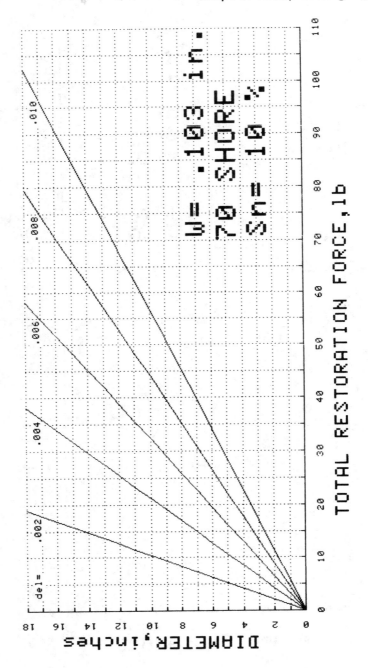

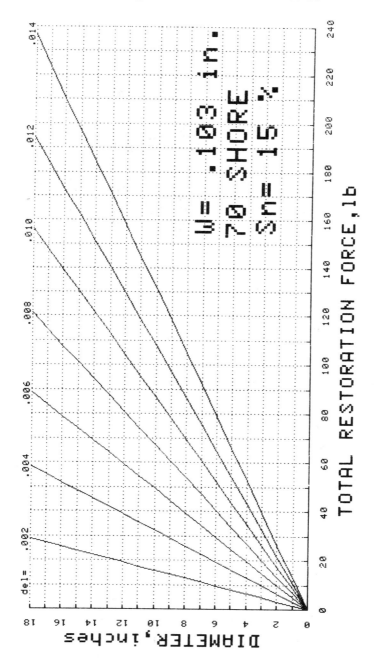

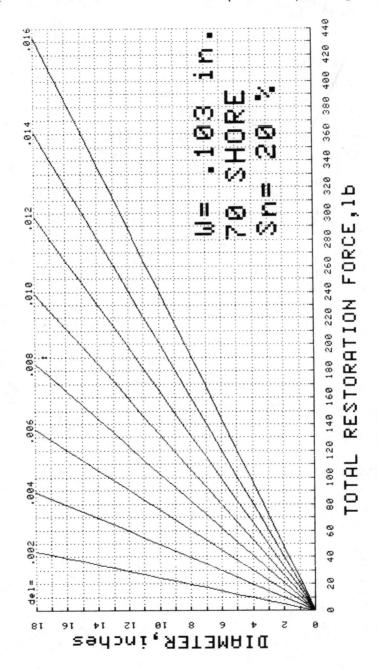

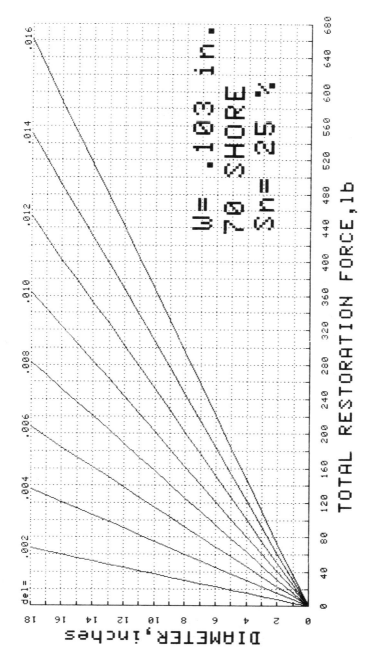

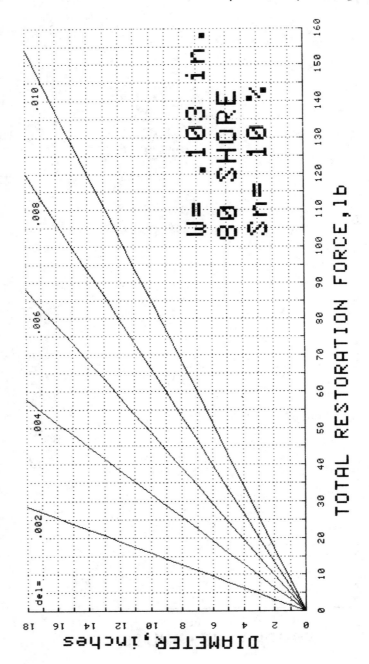

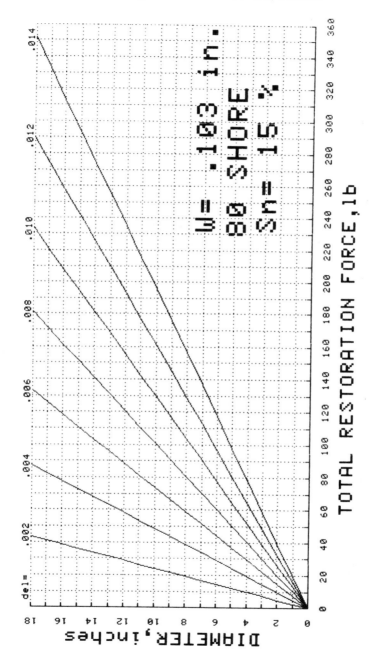

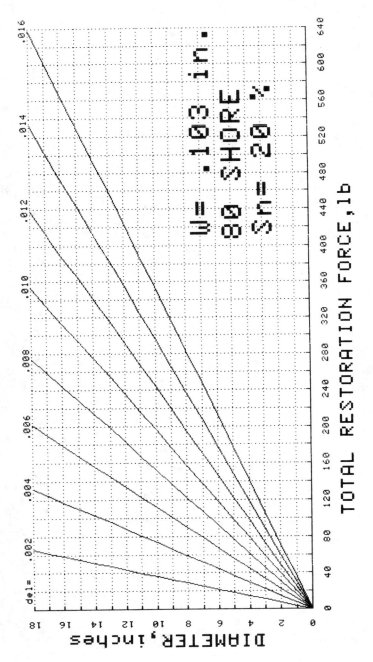

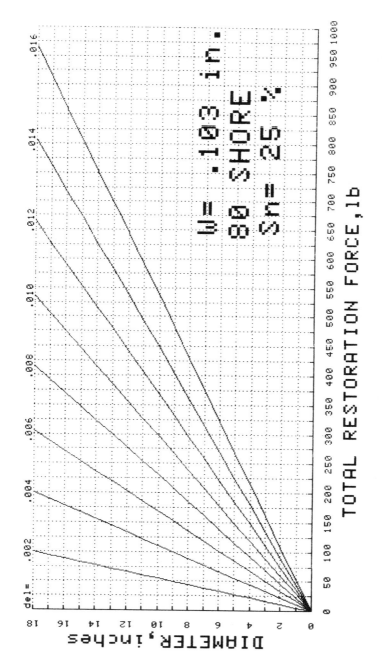

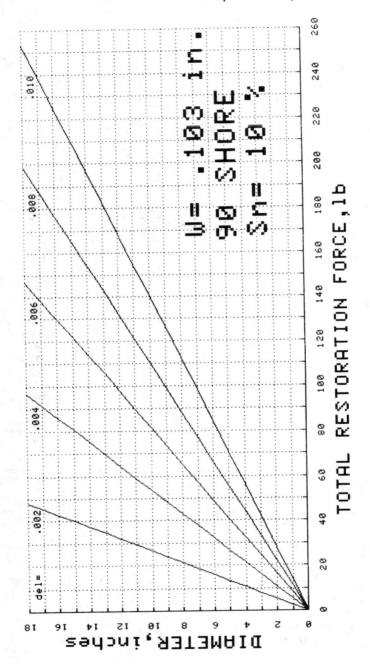

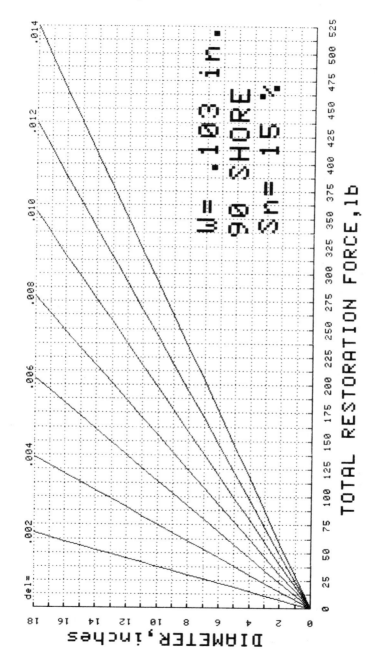

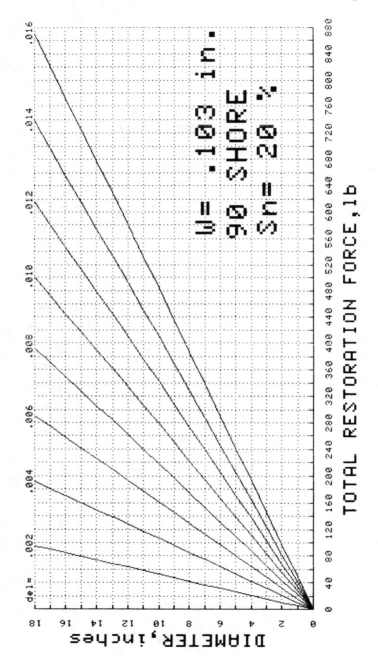

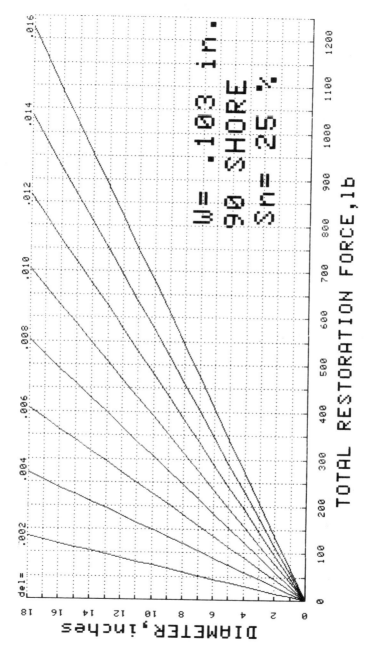

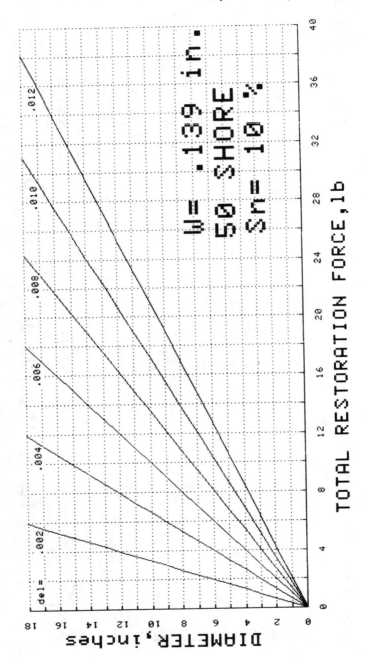

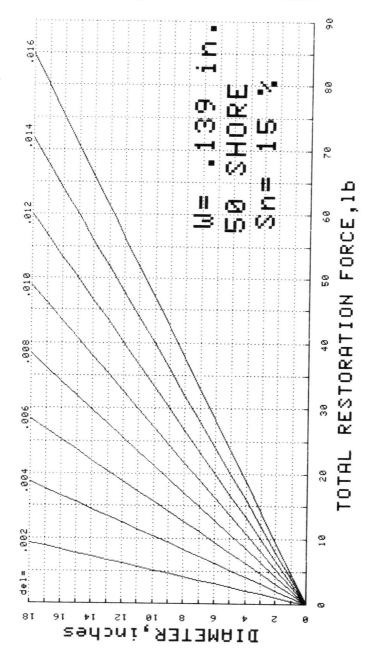

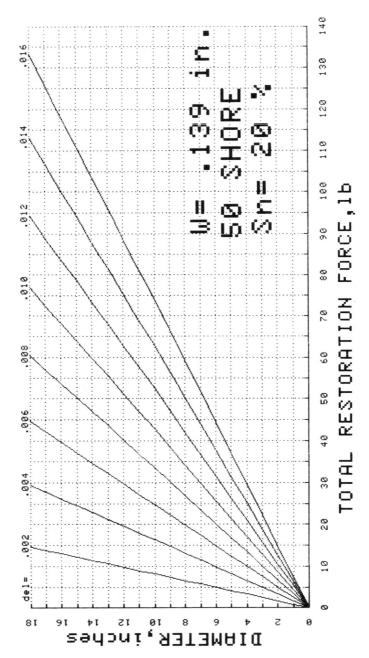

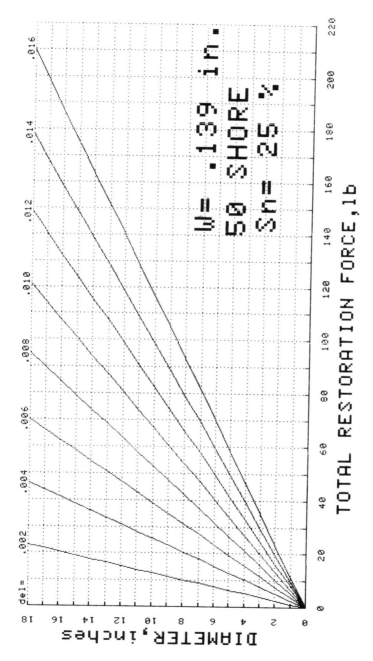

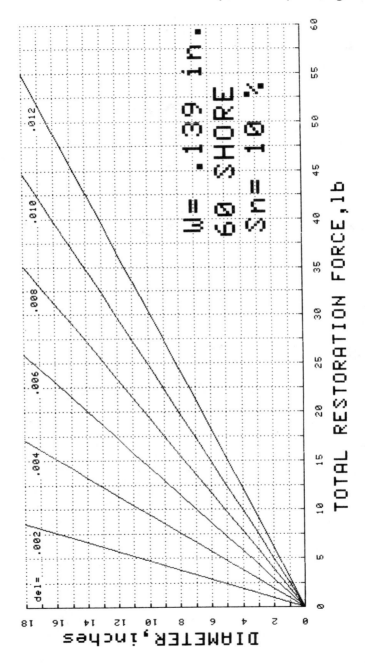

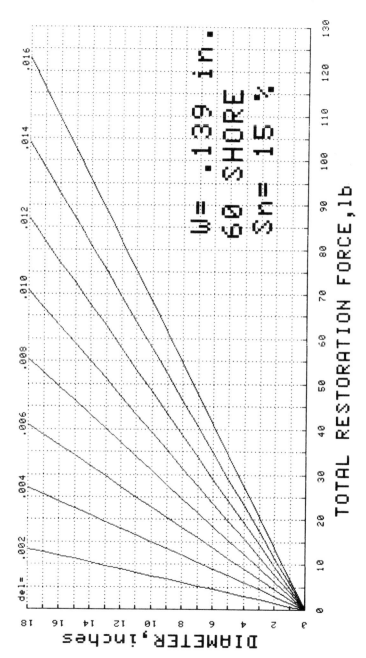

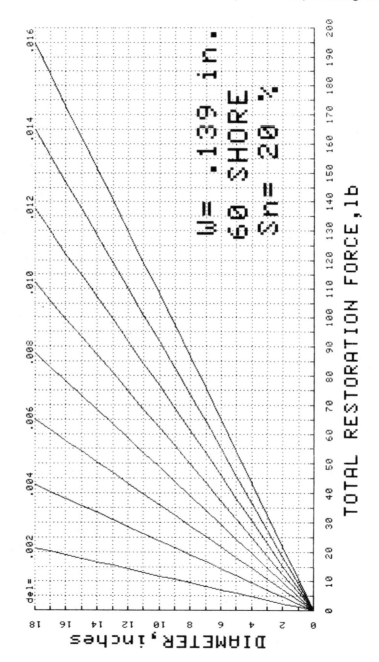

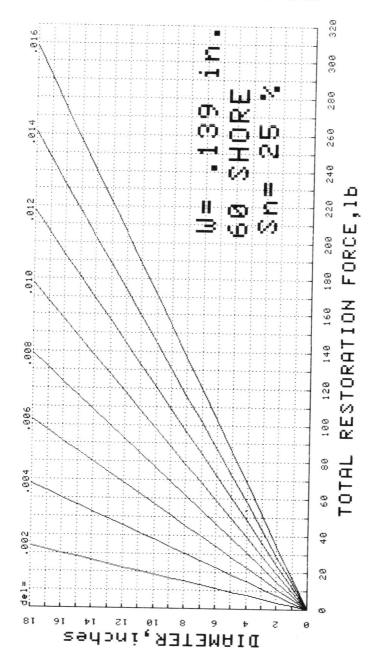

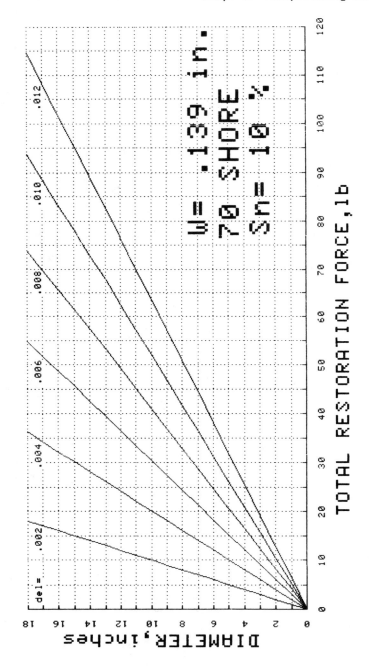

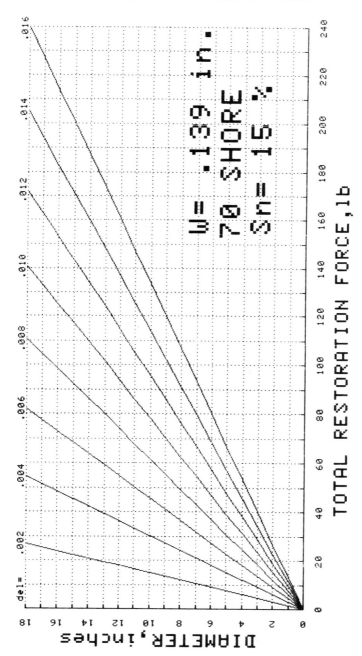

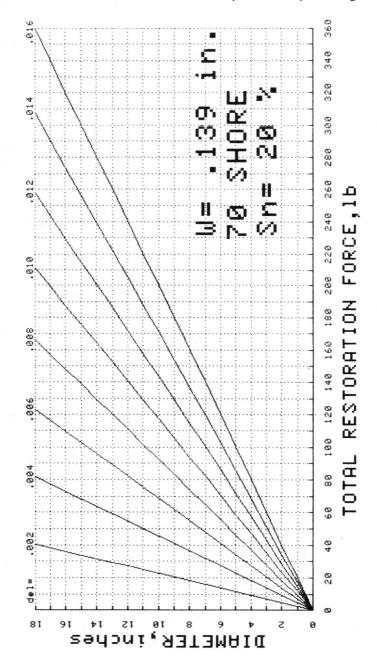

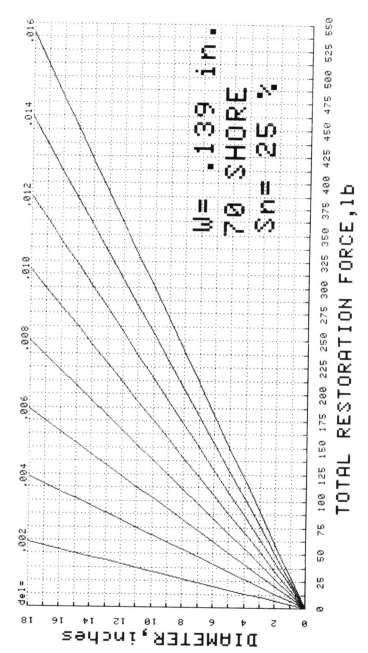

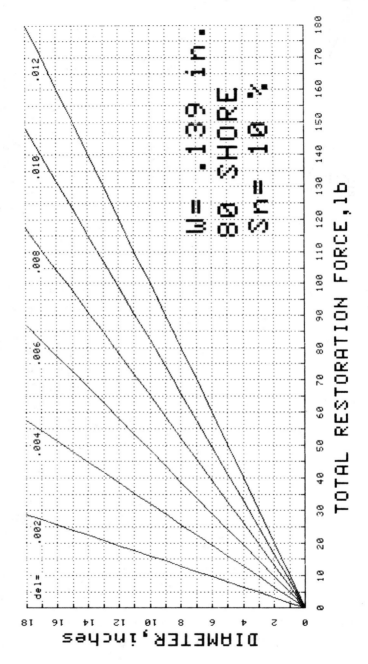

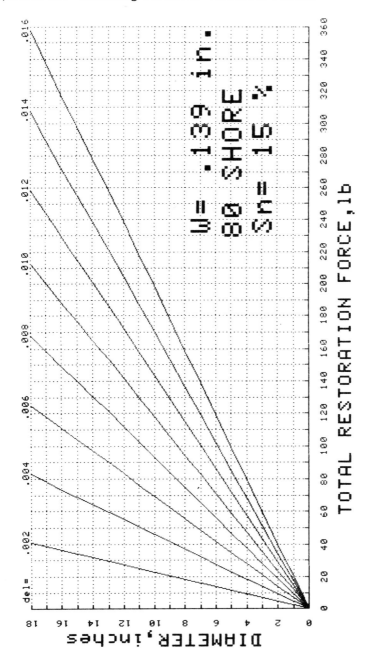

W= .139 in.
80 SHORE
Sn= 15 %

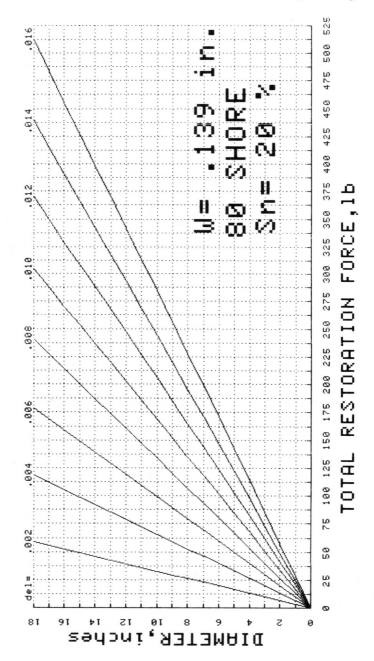

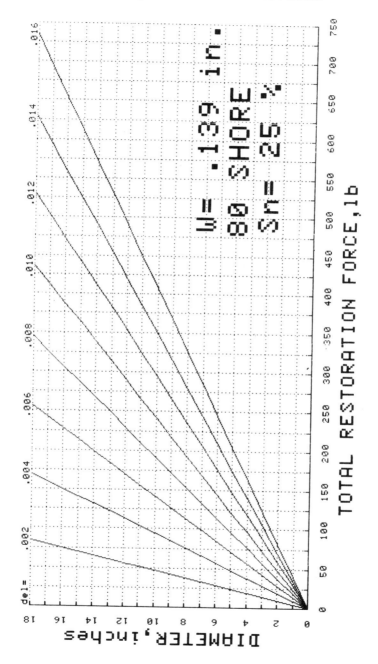

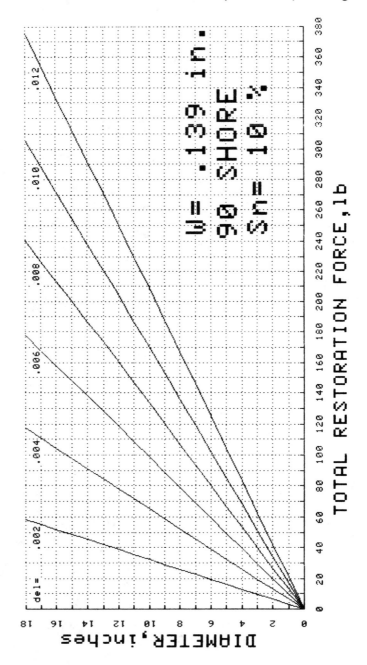

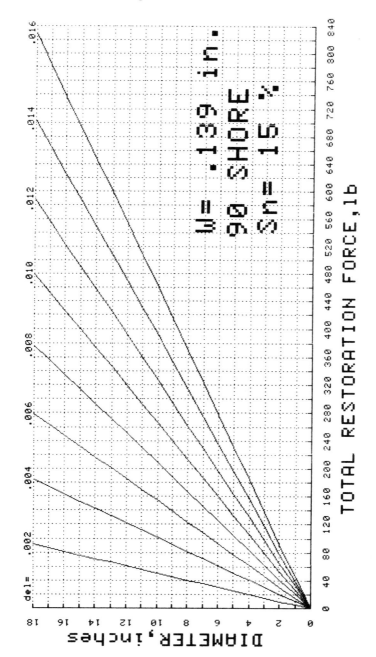

W= .139 in.
90 SHORE
Sn= 15 %

DIAMETER, inches

TOTAL RESTORATION FORCE, lb

del= .002 .004 .006 .008 .010 .012 .014 .016

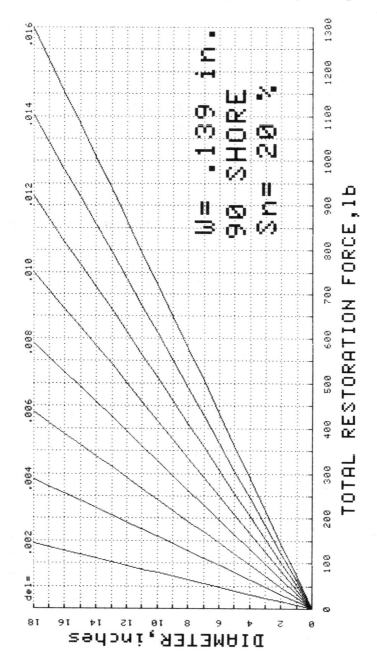

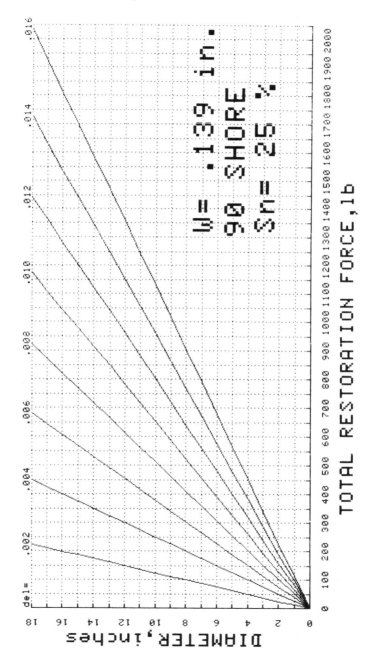

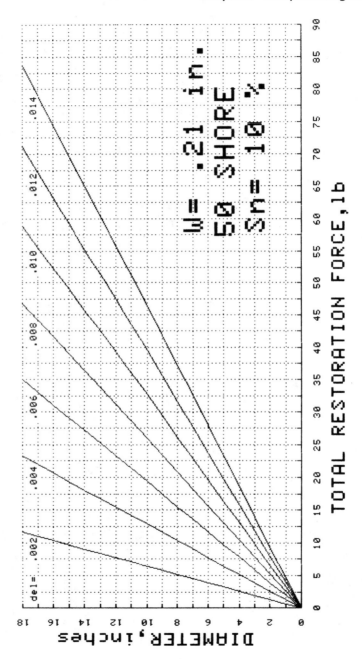

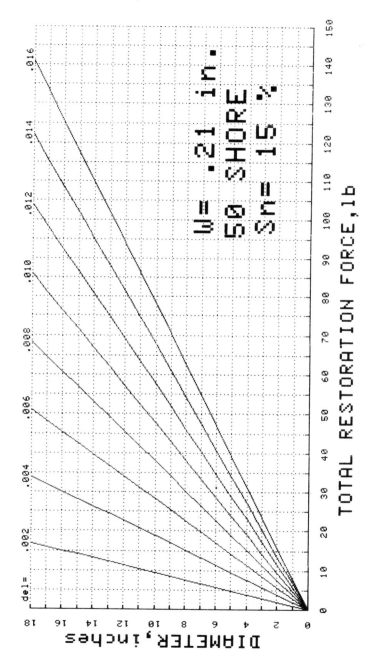

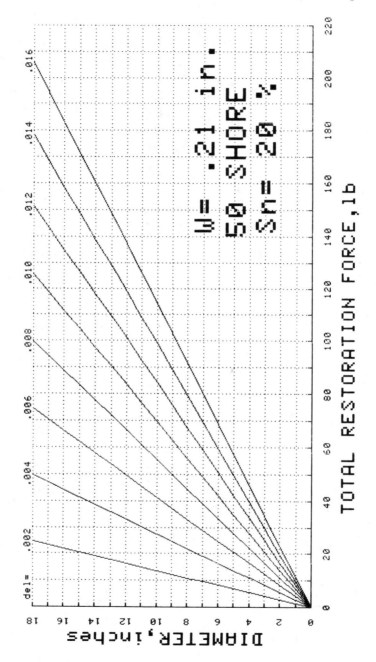

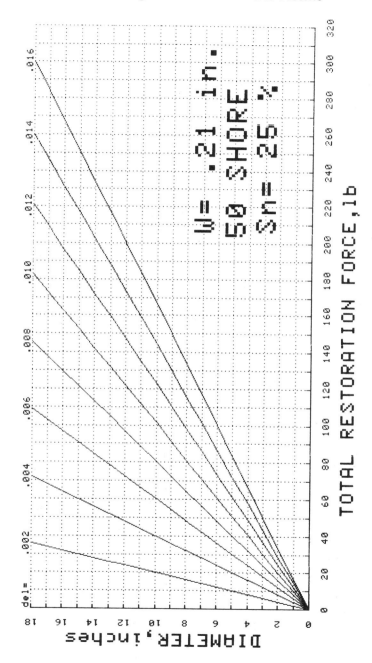

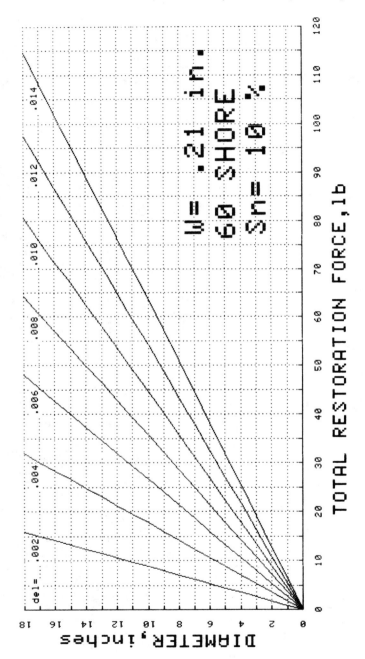

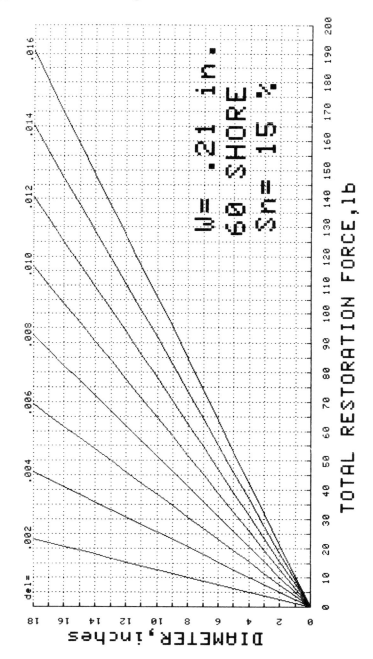

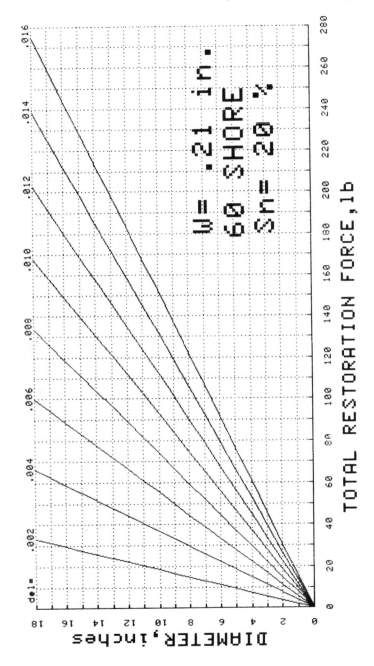

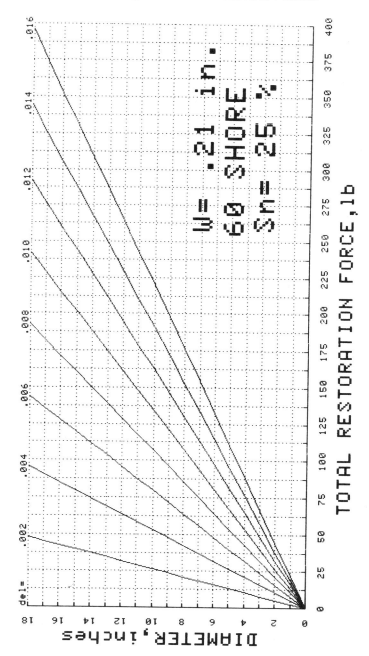

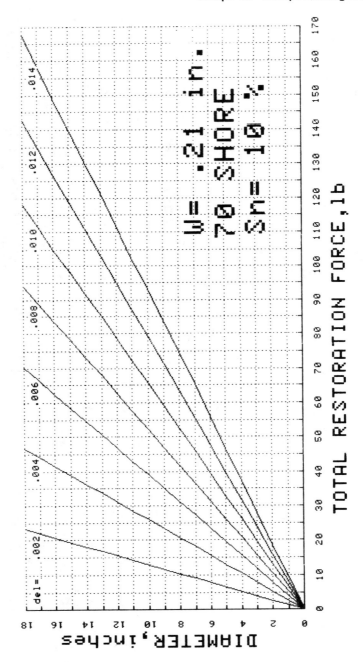

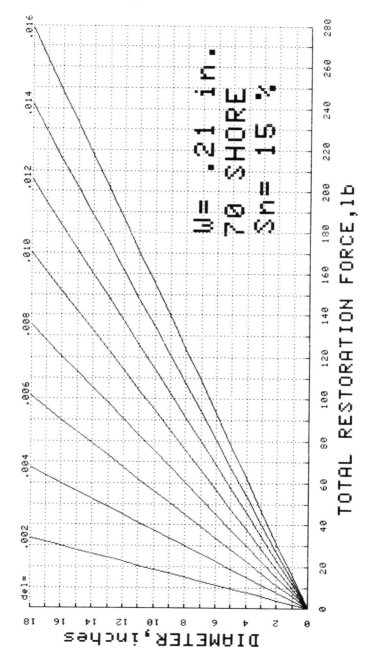

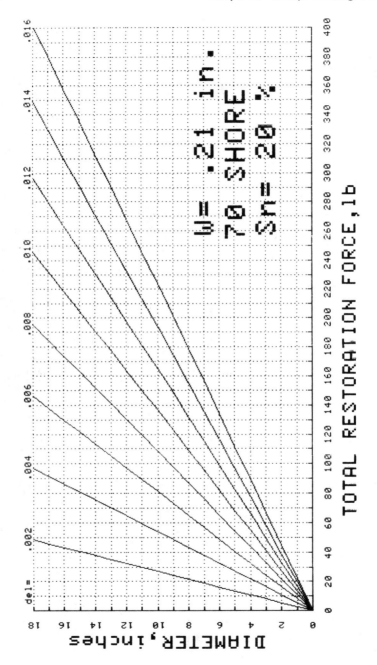

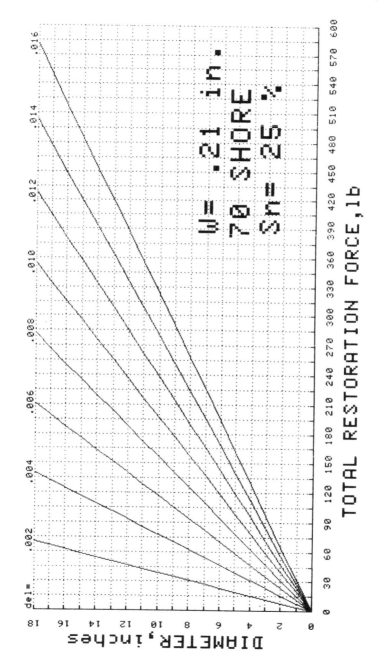

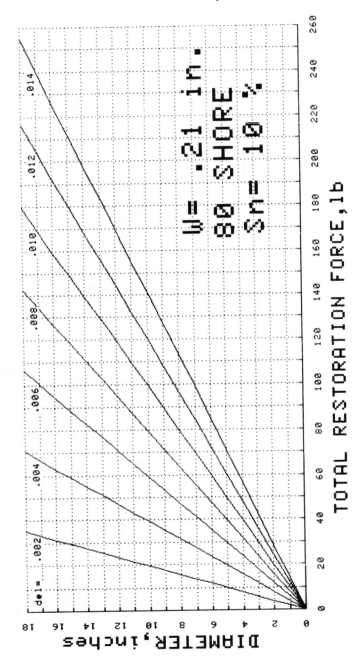

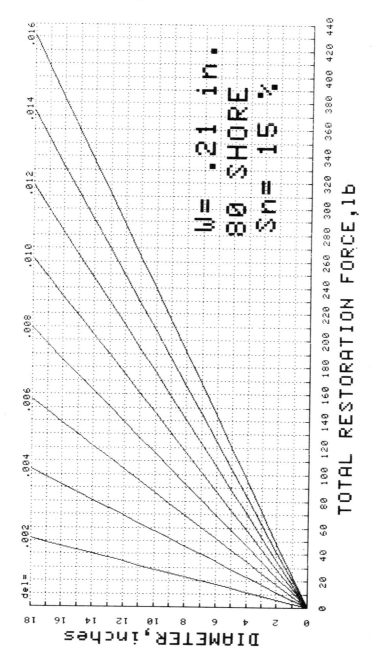

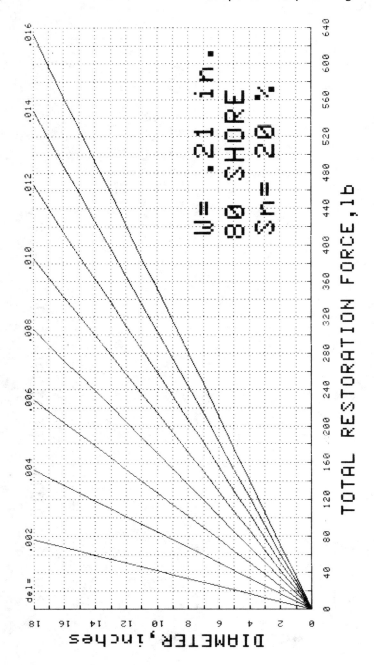

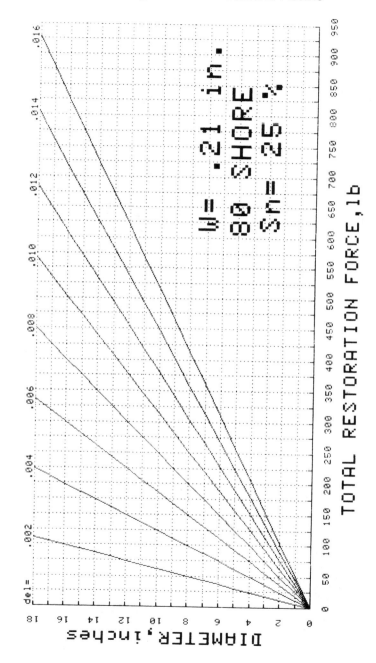

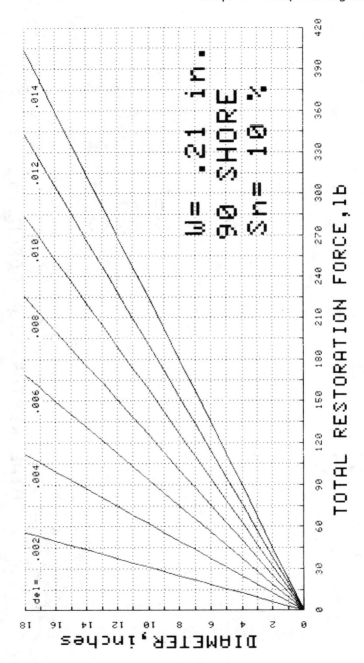

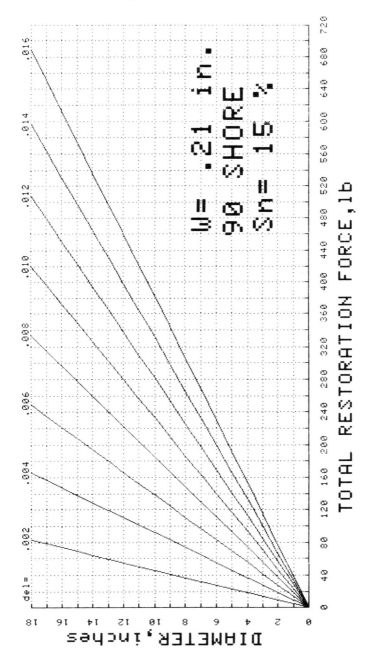

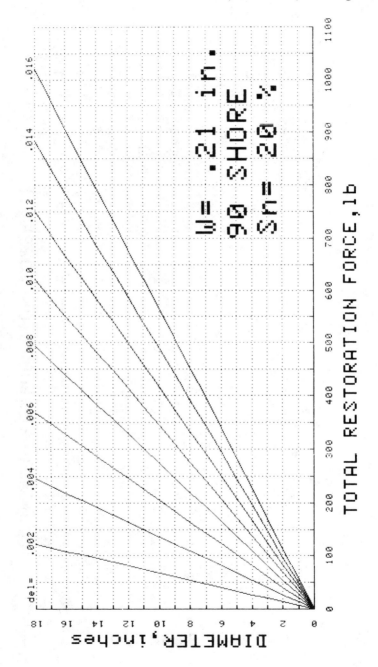

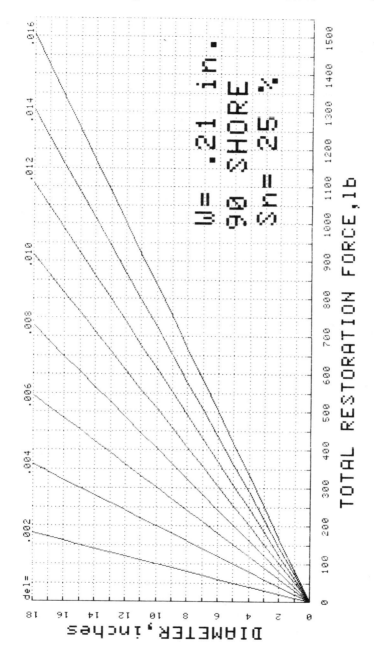

DIAMETER, inches

TOTAL RESTORATION FORCE, lb

W= .21 in.
90 SHORE
Sn= 25%

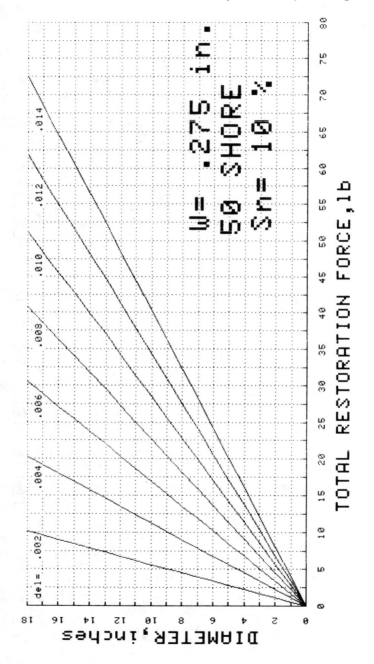

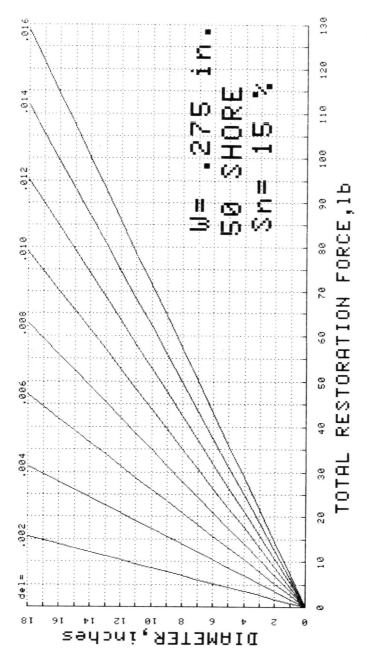

W= .275 in.
50 SHORE
Sn= 15 %

TOTAL RESTORATION FORCE,lb

DIAMETER,inches

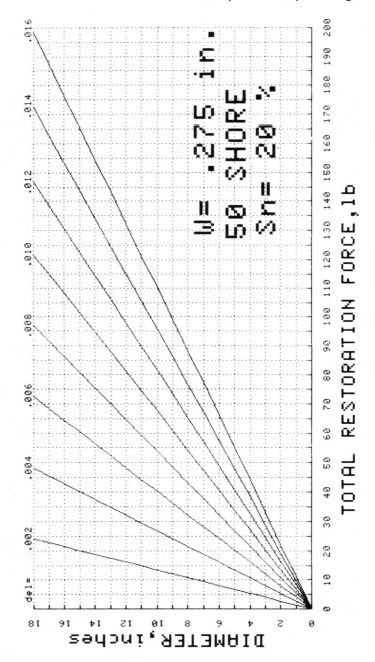

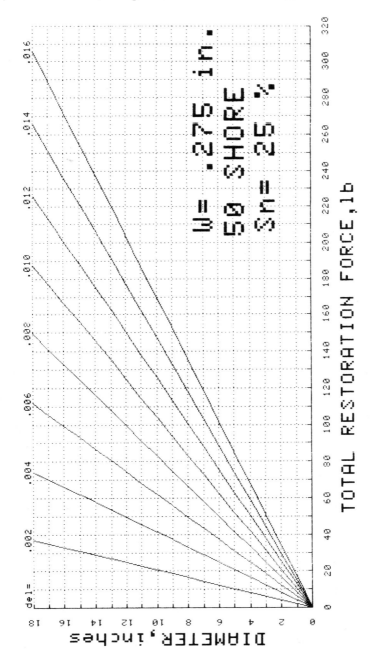

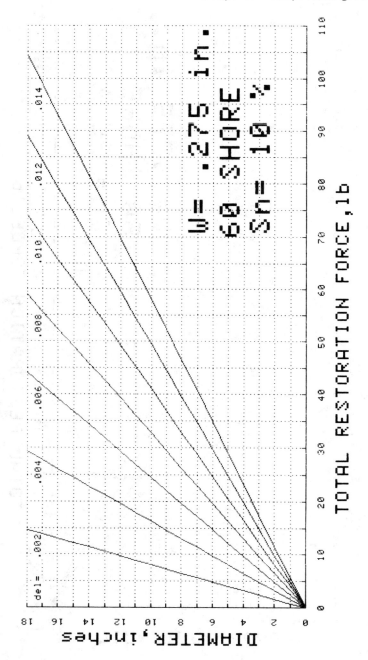

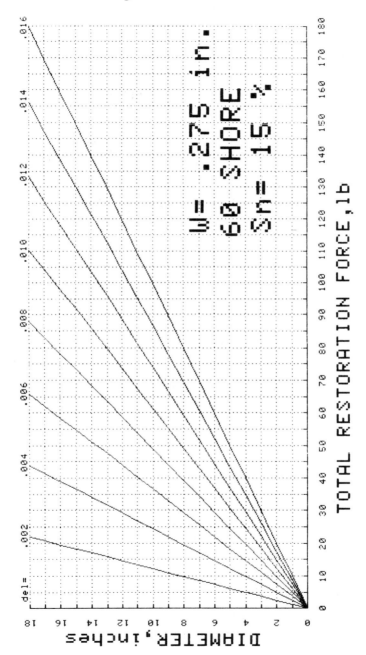

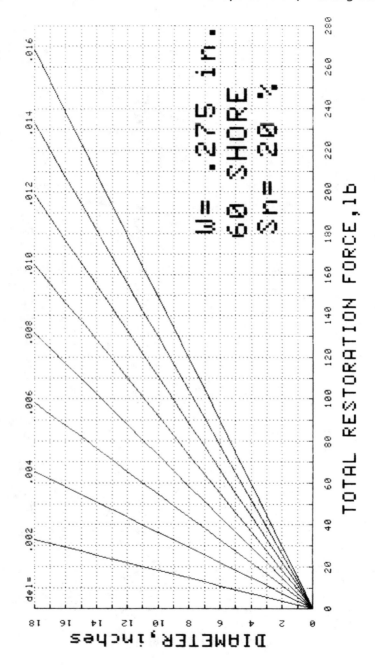

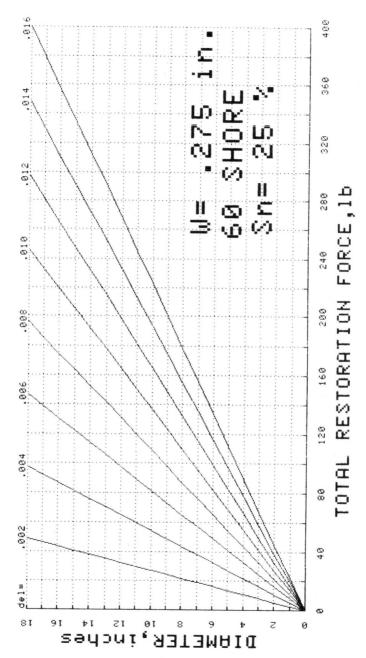

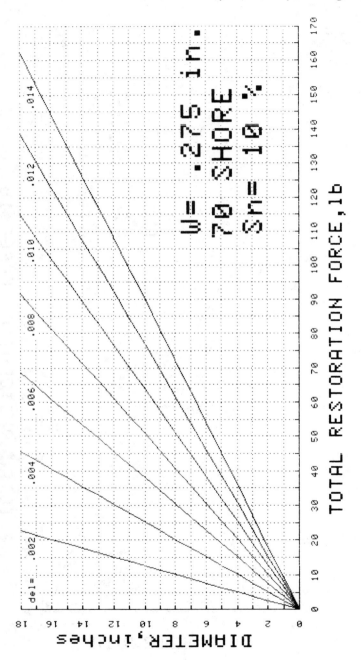

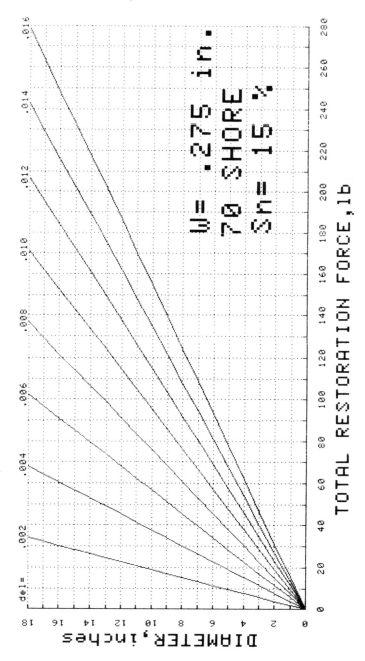

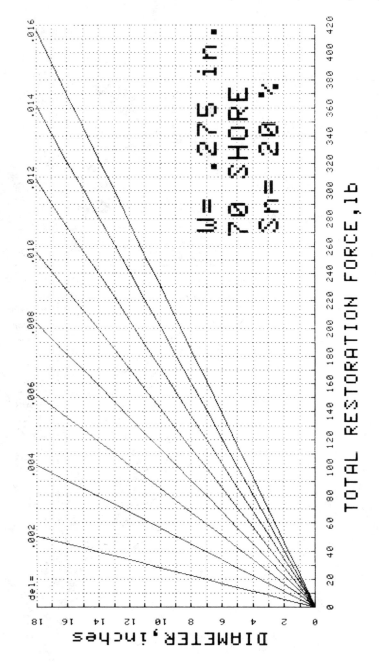

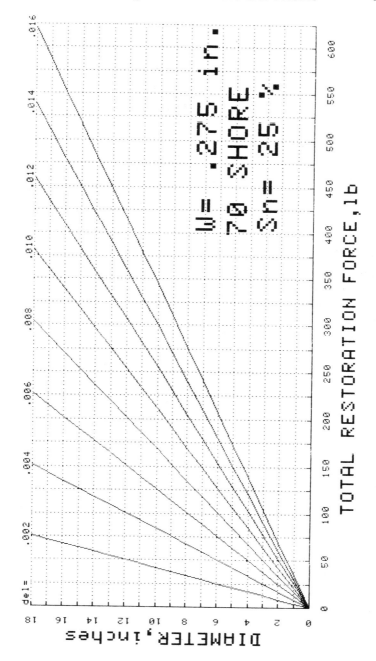

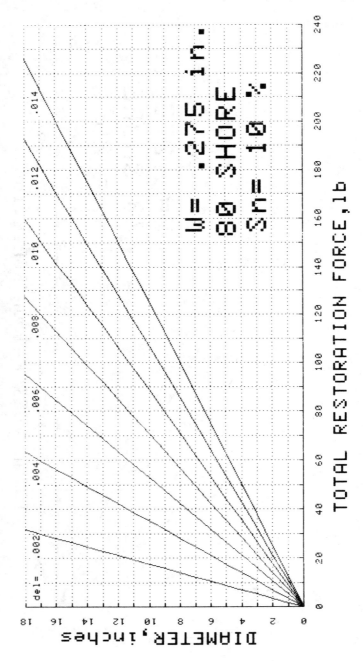

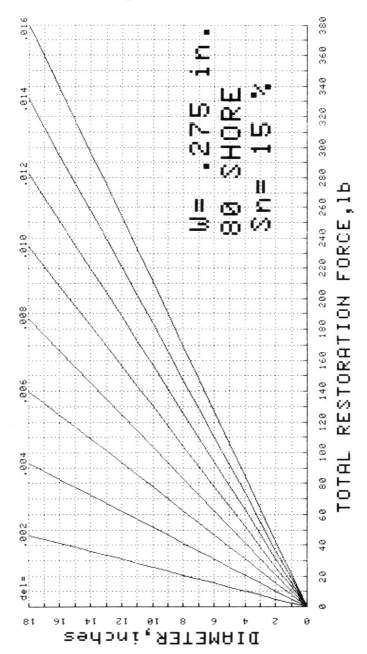

DIAMETER, inches

TOTAL RESTORATION FORCE, lb

W= .275 in.
80 SHORE
Sn= 15 %

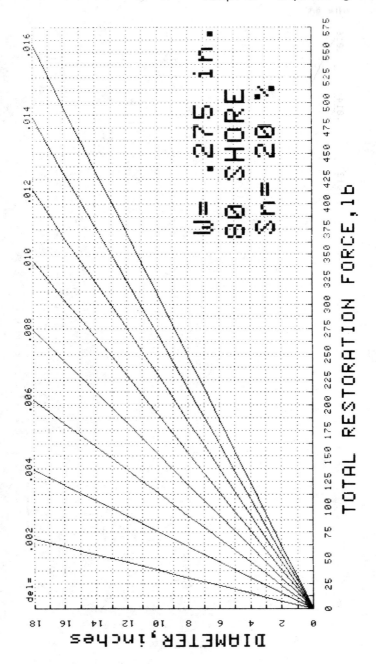

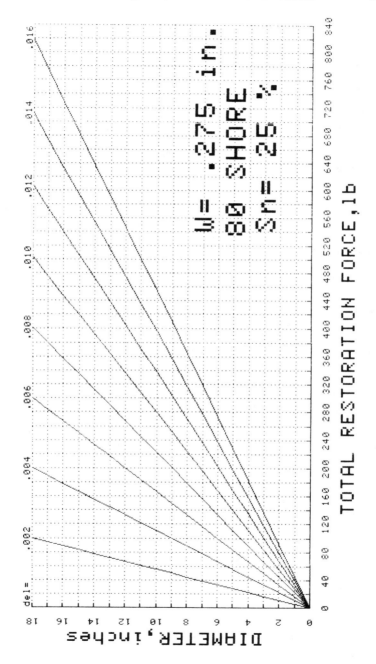

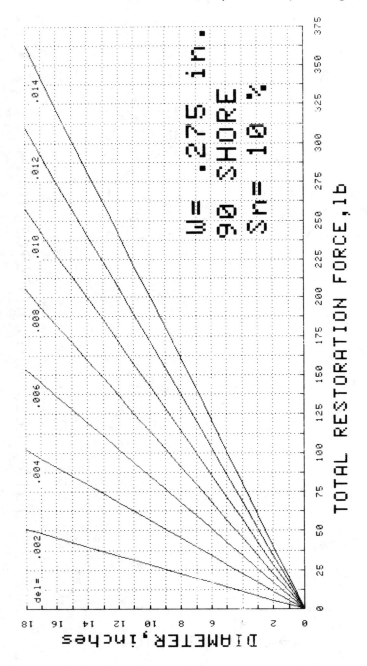

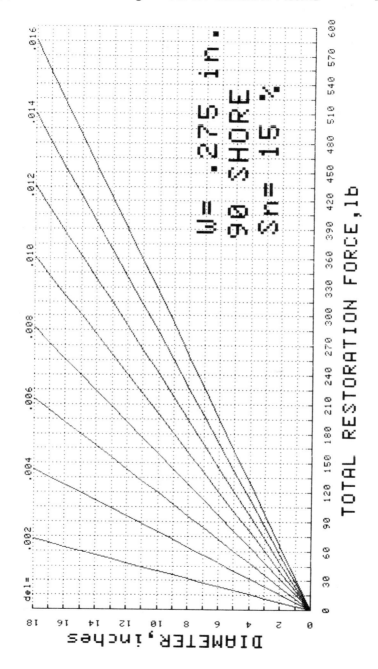

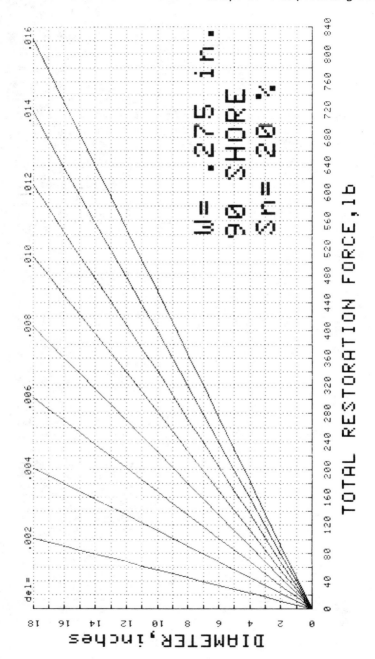

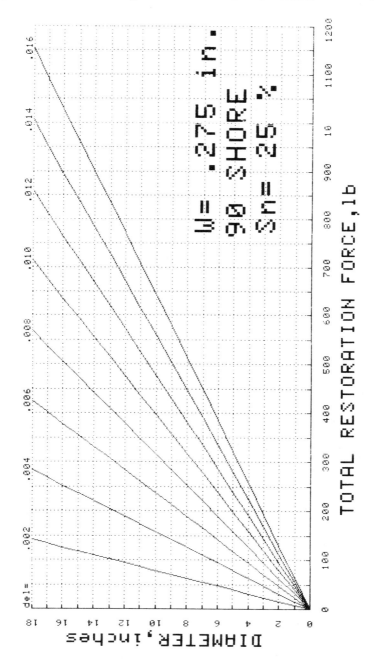

6

Rotary Seals—Designs for Rotating Shafts

I. THEORY*

A. Theory of O-Ring Rotary Sealing Applications

Once it was considered impossible to use O-rings in a seal for a rotating shaft except at very low ratational speeds. This inapplicability of O-rings to rotary motion was blamed on the Gow-Joule effect. According to this phenomenon, when rubber is stretched, then heated, it tries to contract, and its modulus of elasticity and thus stiffness or ability to carry load increases with rise in temperature. If the rubber is under constant load, it will contract; if under constant strain, it will exert greater stress.

When an O-ring is used in the usual way, by stretching slightly around the shaft, the friction of the rubber against the rotating shaft generates heat, which causes the rubber to contract about the shaft. As the rubber O-ring contracts about the shaft, unit loading at the shaft-O-ring sealing surface increases; thus more friction and heat are generated, and the Gow-Joule effect intensifies. The cycle of friction, heat, and contraction of the ring is repeated until rapid failure of the seal occurs.

If the O-ring has peripheral tensile stresses due to its being stretched over the shaft, failure will occur in a few minutes at shaft speeds above 200 fpm [3]. The differential pressure across the O-ring seal also increases the unit loading at the sealing surface and will cause premature failures at even slower shaft speeds.

---

*Part of the material in this section has been reprinted from Leonard J. Martini, *Seals for the Ocean Environment*, Naval Undersea Research and Development Center, San Diego, April 1972, pp. 7-13 and 10-23.

The amount of tensile stress allowable for a rubber seal of any
type that rubs directly on a shaft depends on contact area, differential pressure across the seal, unit loading, surface finish of the
shaft, and the properties of the rubber material used. A design
formula for the relationship between the peripheral tensile stress
and the environmental conditions would be of the form

$$\sigma = \varepsilon E$$

$$= f(A, \Delta P, L, rms)E$$

Notice that the peripheral tensile stress on the rubber O-ring
will be reduced if the contact area and the unit loading caused by
the expansion of the ring on the shaft are reduced. If the reduction of the two latter factors is great enough, the peripheral
tensile stress can be alleviated to the point of peripheral compression; that is, the unit load becomes compressive $(-L)$ instead of
tensile. If the net forces on an O-ring result in peripheral compression even under the severest environmental loads, the Gow-
Joule effect will not occur. This is because the Gow-Joule effect
occurs only if the rubber O-ring is under tensile stress [4].
Therefore, a rotary seal will operate if the induced peripheral
compressive stress is always greater than tensile stresses induced
by differential pressure ΔP and the unit loading L.

Also notice that the peripheral tensile stress σ on the rubber
O-ring will be reduced if the surface finish (rms) of the shaft is
reduced. Theoretically speaking, the static and dynamic frictions
between the O-ring and shaft are caused by microscopic asperities
at the sealing interface. Asperities on the shaft can be reduced
by polishing the shaft. Lubricants also reduce friction by mechanically separating the asperities between the O-ring and shaft
(i.e., effectively reduce the rms surface finish, which in turn
reduces the rate of heat generated, lessening the Gow-Joule
effect).

Figure 21 shows how a rotary shaft seal can be designed to
induce peripheral compression on the O-ring. The dimensions of
the O-ring groove and seal housing are such that when an over-
sized O-ring is installed, the O-ring is forced against the bottom
(maximum depth) of the groove, away from the shaft, as in Fig.
21d. For efficient running conditions, the O-ring should have
about 5 to 8% peripheral compression [(OD − G)/OD] X 100%, and
a cross-sectional squeeze usually between 2 and 4% when the shaft
is installed, [(W' − D)/W'] X 100%. When the O-ring is installed
in the groove (Fig. 21d), the cross-sectional area of the O-ring
increases because its outer diameter is decreased by an amount
OD − G. The O-ring tends to "snake" in the groove, a phenom-

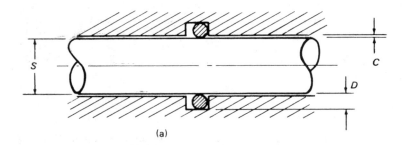

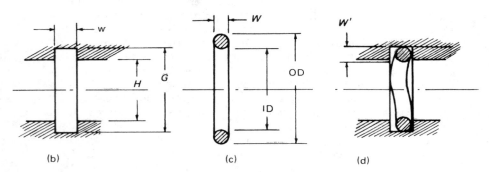

Figure 21. Rotary shaft seal

enon opposing the increase in cross-sectional area. If the amount
of "snake" is excessive, the cross-sectional area of the O-ring will
not increase sufficiently to produce the desired amount of peripheral compression. Therefore, the groove width w is usually
kept to a minimum. Since the groove width must allow for swell
caused by fluid adsorption and thermal expansion caused by heating, there is an allowable minimum value of w such that the
volume of the groove is at least 5% greater than the volume of the
O-ring seal [5].

The environmental parameters on which each of the groove
dimensions depends can best be defined by a relationship table (Table
19) and Fig. 22. Design Example 11 represents an actual design
developed by the Naval Ocean Systems Center. It is given here
to show how the relationships in Table 19 apply to each other.
For instance, the clearance between the shaft and seal housing,
C, should be small enough to ensure against O-ring extrusion by

Table 19. Relationships Between Dimensions and Environmental Parameters

Dimension	Environmental parameter	Interrelationships and functions
Seal housing		
C	Differential pressure ΔP	$C \propto \dfrac{1}{(\Delta P)^n}$ (Fig. 5)
H	Shaft size S Differential pressure ΔP	$H = S + 2C$
D	O-ring cross-sectional diameter W Shaft speed N	$D = \eta W$ $\eta \propto N^n$, $\eta = 0.90$ to 0.95
G	Shaft size S Shaft speed N	$G = S + 2D$
w	O-ring cross-sectional diameter W Shaft speed N	Table 20, Groove Width for O-Ring Cross-section
O-ring		
OD	Shaft size S	$OD = fG$
ID	Shaft speed N	$f = 1 + \dfrac{\%\ \text{diametral reduction}}{100\%}$ $ID = OD - 2W$
W	Shaft speed N	$W \propto \dfrac{1}{N^n}$, design chart or minimum W
W'	Shaft speed N	$W' = IW_{min}$ $I = 1 + \dfrac{\%\ \text{increase in cross section}}{100\%}$
f I	Shaft speed N and Differential pressure ΔP	Figure 23, % diametral reduction vs. % increase in cross section

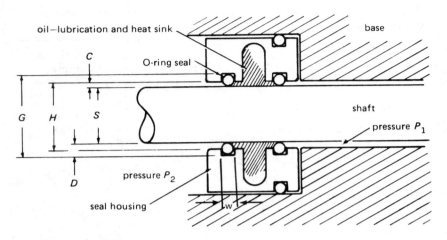

Figure 22. Spindle design—rotary O-ring seals

differential pressure but large enough to allow the seal housing to
"float" and take up eccentricities without contacting the shaft.
The clearance is determined from nonextrusion data such as those
shown in Fig. 5. In the design example, the clearance chosen
was slightly greater than that given in Fig. 5, that is, C = 0.0025
in. instead of 0.0019 in., thus increasing the effectiveness of the
floating housing.

Design Example 11 Rotary Spindle Design

Environmental parameters
Given (referring to Fig. 22):

shaft diameter S = 2.0625 ± 0.0005 in.

shaft speed N = 2200 rpm

differential pressure $\Delta P = P_2 - P_1$ = 2000 psi

Component dimensions
Determine:

1. C, clearance
2. H, housing diameter (± 0.0005 in.)
3. D, gland depth
4. G, groove diameter (± 0.001 in.)

5. O-ring size required, OD and ID
6. O-ring cross-sectional increase in groove
7. Actual squeeze of O-ring
8. w, groove width
9. peripheral compression

Calculations:

1. C, clearance: make maximum allowable that will still ensure against O-ring extrusion by differential pressure ΔP (see Fig. 5).

    C = 0.0025 in. nominal

2. H, housing diameter = S + 2C:

    H = 2.0625 + 2(0.0025) = 2.0675 ± 0.0005 in. diam.

3. D, gland depth: D ≈ (0.90 to 0.95)W, where W = O-ring cross-sectional diameter. Select the smallest W to provide the least sealing area. Use a factor of 0.95 because the shaft speed is high, 1188 fpm (see Table 20). Therefore,

    D = (0.95)(0.070 ± 0.003) = 0.0665 in. nominal depth

4. G, groove diameter = S + 2D:

    G = 2.0625 + (2 X 0.0665) = 2.196 ± 0.001 in. diam.

5. O-ring size required:
    a. OD = f X G, where the factor f is

    $$f = 1 + \frac{\text{\% of diametral reduction}}{100\%}$$

    Usually, f = 1 + 0.08 = 1.08 in., therefore,

    OD = 1.08(2.196) = 2.372 in. diam.

    b. ID = OD − 2W

    $$= 2.372 - 2(0.070) = 2.232 \text{ in. diam.}$$

    Therefore, use the nearest size O-ring:

    ID = 2.239 ± 0.018 in. by W = 0.070 ± 0.003 in. Parker No. 2-35
    OD = 2.379 ± 0.024 in.

6. O-ring cross-sectional increase in groove:
    a. $W'_{min} = I \times W'_{min}$, where the factor I is

    $$I = 1 + \frac{\text{\% increase in cross section}}{100\%}$$

Usually $I = 1.022$ in.; therefore,

$$W_{min} = 1.022(0.070 - 0.003) = 0.0685 \text{ in.}$$

b. $W'_{max} = 1.022(0.070 + 0.003) = 0.0746$ in.

7. Actual squeeze of O-ring:

a. Minimum $= W'_{min} - \dfrac{G_{max} - S_{min}}{2}$

$$= 0.0685 - \frac{2.197 - 2.062}{2}$$

$$= 0.001 \text{ in.}$$

b. Maximum $= W'_{max} - \dfrac{G_{min} - S_{max}}{2}$

$$= 0.0746 - \frac{2.195 - 2.063}{2}$$

$$= 0.0086 \text{ in.}$$

8. W, groove width: According to Table 20, the groove width should be 0.079 in. for an O-ring cross-sectional diameter of $W = 0.070 \pm 0.003$ in. Therefore, specify $W = 0.079 \pm 0.001$ in.

9. Actual peripheral compression:

a. Minimum $= \dfrac{OD_{min} - G_{max}}{OD_{min}} \times 100\%$

$$= \frac{2.355 - 2.197}{2.355} \times 100\%$$

$$= 6.7\%$$

b. Maximum $= \dfrac{OD_{max} - G_{min}}{OD_{max}} \times 100\%$

$$= \frac{2.403 - 2.195}{2.403} \times 100\%$$

$$= 8.7\%$$

The gland depth is a function of the cross-sectional diameter of the O-ring chosen to provide the least sealing area around the shaft. The cross-sectional diameter or O-ring width W can be selected according to shaft speed (see Table 20). As differential pressure across an O-ring increases, the O-ring deforms more and the sealing area around the shaft increases. Therefore, the

Table 20. Allowable Shaft Speed, Gland Depth, and Groove Width
for O-Ring Cross-Sectional Diameter

O-ring cross-sectional diameter W (in.)	Allowable shaft speed N (fpm)	Approximate gland depth D (in.)	Approximate groove width w (in.)
0.139 ± 0.004	Less than 400	0.134	0.155
0.103 ± 0.003	Less than 600	0.098	0.115
0.070 ± 0.003	Greater than 600	0.066	0.079

minimum O-ring width should be used for high differential pressures. The O-ring width is then multiplied by a compressive
stress factor η to determine the gland depth D. This factor determines the amount of compressive stress induced within the O-
ring and is a function of the shaft speed. The greater the shaft
speed, the greater this factor should be, because high shaft
speeds develop more heat at the sealing surface area, and thus
more heat must be removed by the lubricant. This is discussed
more fully in Sec. I.C.

The smallest O-ring width, 0.070 in., should be chosen whenever differential pressures are greater than 1000 psi. The
groove depth D is determined by multiplying the O-ring width
by a compressive stress factor η, usually between 0.90 and 0.95%.
The groove diameter G is determined by adding twice the groove
depth D to the shaft diameter S. After the correct O-ring width
has been determined, the outside diameter OD can be selected.
According to Table 19, the O-ring OD is a multiple of the groove
diameter G and based on a desired diametral compression. As
stated earlier, the ideal diametral (or peripheral) compression is
about 5 to 8%, so the O-ring OD should be approximately 1.08G.

Parts 5 and 6 of the calculations in Design Example 11 contain
factors f and I that are related to the compressive stress factor
η used in part 3. The theoretical interrelationship of f and I is
depicted in Fig. 23. The equations used to determine this plot
appear in Appendix 6A. Thus, as indicated in Fig. 23, and 8%
reduction in O-ring diameter would result in a cross-sectional
diameter increase of between 2 and 4%. The factor used in the
problem contains a 2.2% increase in cross-sectional diameter, the
probable minimum O-ring width when installed in the groove.
Using the minimum percentage increase in cross-sectional diameter
ensures that the O-ring selected will provide a positive minimum
squeeze even when the shaft is not rotating, although the actual

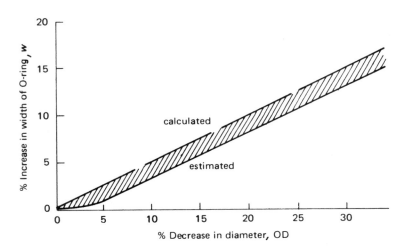

Figure 23.   Theoretical relationship of f and I

squeeze should be checked as calculated in part 7 of Design Example 11.

B. Detailed Theory—Applied and Reaction Forces on Rotary Seals

The determination of the most efficient I factor (percent increase in cross-sectional diameter) depends primarily on one environmental factor: the amount of heat encountered by the O-ring seal. Heat is generated at the sealing surface of the rotating shaft; it may be transferred from the fluid being sealed, from closely located bearings that generate heat, or through the base or seal housing, depending on the application. The heat generated at the sealing surface of the rotating shaft is usually the most critical since it is caused by localized friction produced by relative motion and intensified by differential pressure forces across the O-ring seal. The most efficient I factor must result in a seal with the maximum amount of peripheral compression to oppose tensile stresses which give rise to the Gow-Joule effect. These relationships are shown in Figs. 24a and 24b, which depict the stresses applied to the rotary O-ring and the resulting cumulative stresses within the O-ring. The applied stresses must be balanced by the resultant stresses within the O-ring. The magnitude and direction of the stress are depicted by the length and orientation of the vectors, respectively.

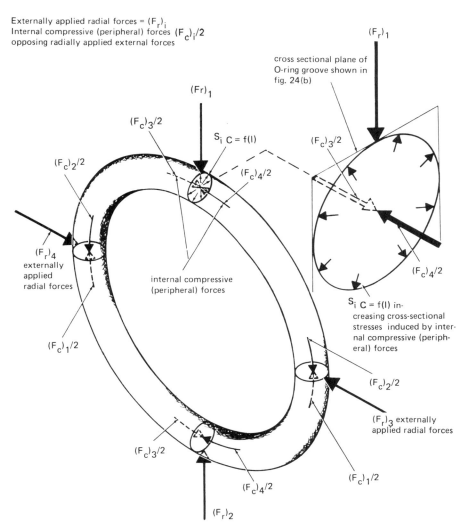

Externally applied radial forces = $(F_r)_i$
Internal compressive (peripheral) forces $(F_c)_i/2$
opposing radially applied external forces

cross sectional plane of
O-ring groove shown in
fig. 24(b)

$(F_r)_1$

$(Fr)_1$

$(F_c)_3/2$

S_i C = f(l)

$(F_c)_2/2$

$(F_c)_3/2$

$(F_c)_4/2$

$(F_r)_4$
externally
applied
radial forces

internal compressive
(peripheral) forces

$(F_c)_4/2$

S_i C = f(l) in-
creasing cross-sectional
stresses induced by internal compressive (periph-
eral) forces

$(F_c)_1/2$

$(F_c)_2/2$

$(F_r)_3$ externally
applied radial forces

$(F_c)_3/2$

$(F_c)_1/2$

$(F_c)_4/2$

$(F_r)_2$

Note: The internal compressive (peripheral) forces at the plane oppose the external radial forces applied 90°
from the plane shown, i.e., $(F_c)_3/2$ opposes $(F_r)_3$.

Figure 24. (a) O-ring in peripheral compression;

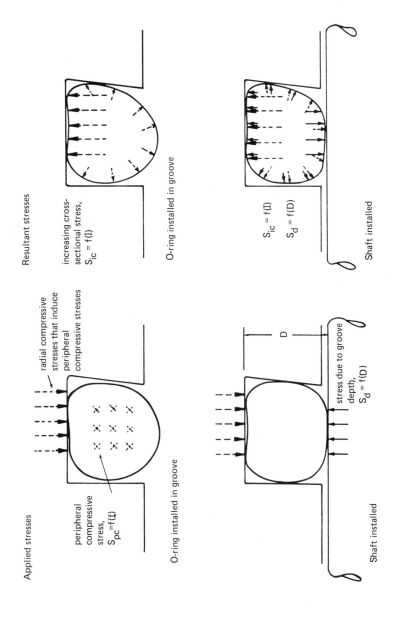

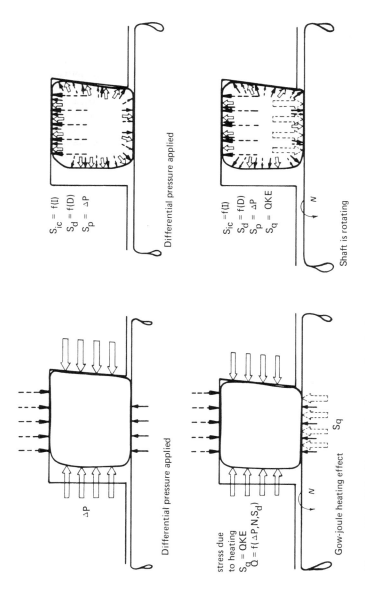

$S_{ic} = f(I)$
$S_d = f(D)$
$S_p = \Delta P$

Differential pressure applied

$S_{ic} = f(I)$
$S_d = f(D)$
$S_p = \Delta P$
$S_q = QKE$

Shaft is rotating

ΔP

Differential pressure applied

stress due
to heating
$S_q = QKE$
$Q = f(\Delta P, N, S_d)$

S_q

Gow-joule heating effect

Figure 24. (b) Rotary O-ring seal—design stresses

Figure 24a is a three-dimensional sketch of an O-ring in peripheral compression. The peripheral compression acts circumferentially within the O-ring to oppose the externally applied radial forces caused by the O-ring groove. Therefore, the internal compressive (peripheral) forces shown acting on the cross-sectional plane, $(F_c)_3/2$ and $(F_c)_4/2$, oppose half of the externally applied radial forces $(F_r)_3$ and $(F_r)_4$, respectively. Even though the cross-sectional plane represents an infinitesimal section of the O-ring, the entire O-ring volume is in peripheral compression. The peripheral compressive forces, or stress, induce cross-sectional stress S_{ic} in the plane. These induced cross-sectional stresses actually increase the cross-sectional area of the O-ring (phenomenon of Poisson's ratio).

Figure 24b is an elaboration of the applied and induced stresses shown in the cross-sectional plane of the O-ring depicted in Fig. 24a. The first views show the indirectly applied peripheral compressive stress as a function of the percent of diametral reduction f. Applying this stress causes an increase in cross-sectional area (shown in the adjacent view) and a corresponding resultant stress throughout, dependent on I. (The relationship between f and I was shown in Fig. 23.) The second set of views shows the shaft installed, cumulative stresses after the addition of the stress resulting from the depth of the groove being about 95% of the original O-ring width, and the consequent internal stresses applied by the O-ring. The third set of views shows the application and resultant stresses of differential pressure (i.e., the function of the O-ring to seal a fluid under pressure). The stresses resulting from both groove depth and differential pressure are equally distributed inside the O-ring.

This equalized distribution follows from the fact that a rubber seal is considered to be an incompressible, viscous fluid having a very high surface tension [4]. Although these two stresses are distributed evenly within the O-ring seal, they act to flatten the O-ring against the shaft, thereby increasing the contact area of the seal and the unit loading on the shaft. This increase in unit loading increases the frictional forces as the shaft rotates and would produce a Gow-Joule effect if not compensated by the opposing stress (S_{ic}) caused by increased cross section.

The Gow-Joule effect is shown in the last set of views only to depict what the S_{ic} opposes and to indicate that heating effects Q are a function of the differential pressure, groove depth, and shaft speed. It must be remembered that as long as the net stresses inside the O-ring are opposed by the initially induced peripheral compressive stress S_{pc}, the Gow-Joule heating effect cannot occur and the O-ring will not even start to contract around the rotating shaft.

In the last view, it becomes obvious that the S_{ic} stresses should be made maximum to oppose the stresses that would otherwise give rise to the Gow-Joule effect. To obtain this maximum cross-sectional stress, the maximum amount of peripheral compression f must be applied to the O-ring. Theoretically, this can be accomplished by pushing an O-ring with the largest possible outside diameter into the groove. The limiting factor is the mechanical difficulty in compressing a large O-ring into a small groove without detrimental effects, such as "snaking," scratching, and so on, although O-ring insertion tools, such as the one shown in Fig. 25, greatly ease O-ring installation while preventing O-ring damage. Realistic maximum values of f vary from 1.07 to 1.10 (7 to 10% of diametral reduction). This means that the maximum possible percentage increase in cross-sectional width is 5% (see Fig. 23).

This tool, which operates somewhat like a syringe, makes insertion of O-rings into rotary glands and deep grooves a simple job. Thus special assemblies with access ports are not required. The tool consists of a disklike mandrel, a stem, an inside cavity, and a plunger. In operation, the O-ring is pulled over the mandrel and positioned around the stem in the cavity. The tool is then installed in the bore so that the mandrel is positioned at the O-ring groove. When the plunger is pushed forward it forces the O-ring into the groove [6].

Figure 25. O-ring insertion tool (courtesy of U.S. Naval Ocean Systems Center, San Diego, Calif.)

Another type of O-ring insertion tool consists of a tube with a conical bore which can be positioned at the edge of the O-ring groove within the seal housing. A plunger is then used to push the O-ring down the conical bore toward the O-ring groove. The conical bore reduces the outside diameter of the O-ring such that it slips easily into the O-ring groove. Lightly greasing the O-ring usually helps installation and tends to prevent twisting and excessive distortion of the O-ring.

C. Heating and Lubrication

The surprising feature of elastomeric O-rings when used in rotary sealing applications is their ability to wear metal parts. Friction between an O-ring and any bearing surface is caused by the asperities, or microscopic hills and valleys, in the surface interface. Even though an elastomeric O-ring is much softer than the metal surface sliding against it, the O-ring will wear away the asperities of the metal as it glides past its irregular surface. The result is usually a polishing effect on the metal surface, but if the contact pressure and surface speed at the interface is great enough, excess wear to the point of leakage may occur. We have already seen how peripheral compression can induce cross-sectional stresses within the O-ring that combat the Gow-Joule effect. Lubricants can also combat the Gow-Joule effect by mechanically separating the surface asperities of the O-ring and shaft.

As noted, the amount of heat generated at the sealing surface of the moving shaft and stationary O-ring is increased by an increase in O-ring squeeze (i.e., a decrease in the gland depth D) (Fig. 22). It was also noted that an increase in differential pressure applied across the O-ring increases the amount of heat generated. This heat must be dissipated to prevent overheating of the O-ring and the metal surface of the rotating shaft. This is accomplished through lubrication. The lubricant shown in the annulus of the spindle design (Fig. 22) thus provides a heat sink while reducing that portion of heat generated by normal asperity interference.

The importance of lubrication, especially in an application of high shaft speed and high differential pressure, cannot be overemphasized. The extreme pressure (EP) oil used in the rotary seal shown in Fig. 22 is a special formula of Wynn oil. High-load-carrying films from EP oils have a "cushioning effect" on the normal stresses of asperities. This is accomplished by the EP oil maintaining a sufficiently thick solidlike boundary film, which has a lower elastic modulus than the asperities that this film separates [5].

In addition to mechanically separating asperities between bearing surfaces, boundary lubricants function through chemical phenomena. There are two principal chemical reactions that must be controlled in order to optimize the benefits from a boundary lubricant: (1) beneficial decrease in fracture properties of metal surfaces, and (2) detrimental increase in corrosive wear. A properly balanced boundary lubricant contains oxygen, water, and load-carrying additives that control the amount or rate of conversion of bearing metals to friable corrosion products. These friable products will break into smaller wear particles before causing extensive damage to the confining surfaces. The detrimental effect occurs when too much friability leads to corrosive wear. Therefore, a trade-off of boundary film properties is required [5].

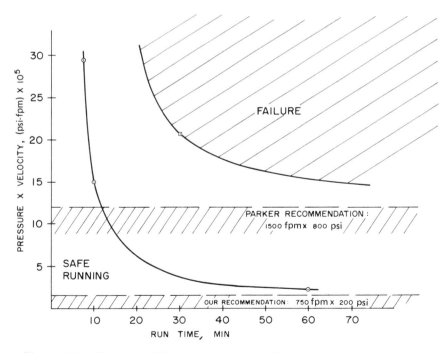

Figure 26. Extreme PV values versus safe running time

D. When to Use This Rotary O-Ring Design

The design for peripheral compression of the O-ring should be
used in every rotary O-ring seal application regardless of shaft
speed or differential pressure. Remember that O-ring failure will
not occur by the Gow-Joule effect if the net forces on the O-ring
result in peripheral compression even under the severest environmental loads. Therefore, designing all rotary O-ring seals for
peripheral compression increases the safety factor in which the
induced peripheral compression stress is always greater than the
tensile stresses induced by differential pressure and the heat
generated by shaft rotation.

In order to predict service life of O-rings used in this peripheral compression rotary seal design, the PV parameter is plotted
against run time. The PV parameter is the product of the pressure exerted on the O-ring in psi and the shaft velocity in fpm.
Figure 26 shows extreme environmental conditions of PV greater
than 100,000 psi-fpm. As can be seen, safe running time becomes limited when PV conditions are greater than 150,000 psi-
fpm, while conditions less than this will not produce a Gow-Joule
failure. The "failure" curve is based on the same slope as the
"safe running" curve and intercepts the only test failure that
occurred. The 1500 fpm and 800 psi (PV = 12×10^5 psi-fpm)
recommendation of the Parker Seal Company is included in the
figure to show that the area between the "safe running" and
"failure" zones is an acceptable condition if other parameters such
as shaft hardness, surface roughness, lubrication, O-ring materials, and concentricity between parts are stringently specified.

The designer must realize that although it is always beneficial
from a service life standpoint to use the peripheral compression
design, it does complicate O-ring installation. O-ring installation
is complicated by the fact that the peripheral compression design
incorporates an O-ring which is 5 to 8% larger in outside diameter
than the diameter of the groove into which it must be installed.
In actuality, this only becomes a design consideration for shaft
sizes less than 2 in. in diameter. The use of O-ring insertion
tools, as discussed in Sec. I.B, alleviates installation problems
associated with designs of smaller size where human fingers seem
too large and cumbersome. For shaft sizes larger than 2 in. in
diameter, O-rings can be eased into place by working the O-ring
around the inside of the groove with the fingers of both hands.
This is discussed under general installation and assembly procedure for the dual-O-ring peripheral compression design presented
in Sec. II.C.

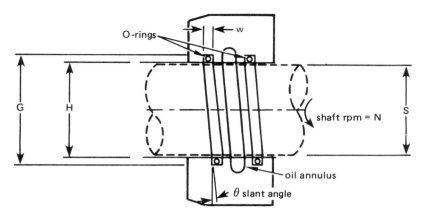

Figure 27. Self-lubricating gland seal—slanted O-ring groove.
Slant angle should be such that the average reciprocation speed
is larger than 20 fpm: $2(S)(\tan\theta)N > 20$ fpm.

E. Self-Lubricating Gland Seal—Slanted O-Ring Grooves

An efficient rotary shaft seal may consist of two O-rings in
peripheral compression and an annulus of oil between them. This
type of design is presented as Design Example 11. If the O-rings
are slanted relative to the axis of the shaft, as in Fig. 27, they
will effectively sweep across its length, bringing new lubricant
into the sealing interface (surface of the shaft under the O-ring
seal) with each revolution. The renewal of fresh lubricant will
greatly reduce the amount of heat created at the O-ring, this
heat being the main cause of the Gow-Joule phenomenon.

Test data for O-rings up to 2 in. in diameter, indicate that
slanting the O-ring seal relative to the rotating shaft not only
extends the life of the shaft but usually reduces the frictional
torque at the seal interface. This is expected because less wear
generally means less running torque. This is reflected in the
empirical results presented in Fig. 28. Notice that the torque for
the slanted groove design initially remains constant and then
gradually decreases to one-third of the torque for the vertical
groove design. This is because the O-ring in the slanted groove
must wear in a larger area of the rotating shaft as it apparently
reciprocates back and forth along the shaft's surface. In so
doing the O-ring also sweeps in fresh oil that acts to lubricate
and slow down this wearing-in process, the ultimate result being

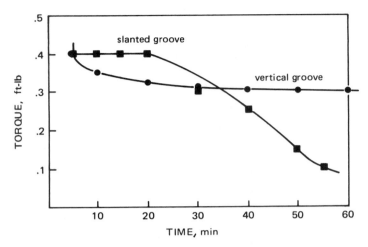

Figure 28. Torque reduction—slanted groove

a distributed polishing effect along the shaft surface instead of the deep wear line as caused by the O-ring in the vertical groove design.

Although a torque-time plot of a rotary seal may seem to predict a reduction in shaft wear, it does not tell the whole story. For as was the case in Fig. 28, one could predict that since the areas under the curves represent a force history, there should have been equivalent wear on each of the shafts. The fact is the wear caused by the vertical O-ring was all at the same point on the shaft, rendering the design incapable after 1 hr, whereas the slanted groove design was still functioning after 15 hr. Additional data also indicate that slanting the O-ring can effectively extend the wearlife of a rotating shaft without apparently reducing the torque required to run it. Evidently, the speed at which the O-ring sweeps back and forth along the shaft's surface is a factor in reducing the frictional torque required to rotate the shaft. Although the reciprocating motion of the O-ring across the surface of the shaft is sinusoidal, consisting of an acceleration and deceleration for every pass, a sort of average reciprocating speed can be defined as the distance the O-ring travels axially along the shaft in one rotation of the shaft; that is, 2(shaft diameter) X (tangent of slant angle) X (shaft speed). For Design Example 12 this reciprocation speed is

$$2(0.563)(\tan 5°) \times \frac{\text{in.}}{\text{rev}} \frac{\text{ft}}{12 \text{ in.}} \times \left(3000 \frac{\text{rev}}{\text{min}}\right) = 24.6 \text{ fpm}$$

Design Example 12 presents a method for checking a slanted O-ring gland design with peripheral compression. The nominal peripheral compression on the O-ring checks out to be 8.4%. When the O-ring is installed in a slanted groove, its perimeter takes on an elliptical shape. The slant angle in such designs is so small that the perimeter is enlarged by only about 0.2% over that of a true circle. This results in such a slight deviation in O-ring cross section that its effect can be ignored, the deviation always being within the tolerance range of manufactured O-rings. The slant angle is checked by calculating the average reciprocating speed of the O-ring on the shaft. In the example, this speed is greater than the minimum limit of 20 fpm.

Design Example 12   Self-Lubricating Gland Seal-Slanted O-Ring Grooves

Given:

The design parameters given in Fig. 29.

Determine:

The feasibility of the design by checking the diametral clearance, groove depth, peripheral compression, the actual squeeze of the O-ring cross section, the average reciprocating speed of the O-ring on the shaft, the frictional torque, and the PV parameter.

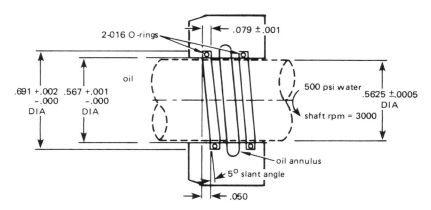

Figure 29.   Self-lubricating gland seal

Design check:

1. 2C, diametral clearance:

$$2C = \left(0.567 \left\{ \begin{array}{l} + \ 0.001 \\ - \ 0.000 \end{array} \right) - (0.5625 \pm 0.0005)\right.$$

$$= 0.0045 \left\{ \begin{array}{l} + \ 0.0015 \\ - \ 0.0005 \end{array} \right. \text{in.}$$

This diametral clearance (0.0060 max.) is well below the maximum
allowable according to Fig. 5 for 500 psi differential pressure.
Therefore, the O-ring will not extrude.

2. D, groove depth:

$$D = \frac{1}{2} \left[\left(0.691 \left\{ \begin{array}{l} + \ 0.002 \\ - \ 0.000 \end{array} \right) - (0.5625 \pm 0.0005) \right] \right.$$

$$= 0.0643 \left\{ \begin{array}{l} + \ 0.0013 \\ - \ 0.0003 \end{array} \right. \text{in.}$$

This is (0.065/0.070) X 100% = 91.4% of the free-state O-ring width.
This is between the recommended range of cross-sectional squeeze
(90 to 95% of W), and is acceptable since the surface speed of the
shaft is low:

$$\pi \left(0.5625 \ X \ \frac{\text{in.}}{12 \ \text{in./ft}}\right)\left(3000 \ \frac{\text{rev}}{\text{min}}\right) = 442 \ \text{fpm}$$

3. O-ring dimensions: From O-ring standard AS568A or Parker
No. 2-016, the dimensions are:

W = 0.070 ± 0.003 in.

ID = 0.614 ± 0.005 in.

4. OD of O-ring:

OD = ID + 2W

= (0.614 ± 0.005) + 0.140 ± 0.006

= 0.754 ± 0.011 in.

5. Nominal diametral compression of O-ring when installed in
groove:

$$\frac{0.754 - 0.691}{0.754} \ X \ 100\% = 8.4\%$$

6. O-ring cross section when installed into groove (the elliptical
perimeter of the O-ring when installed in the slanted groove
deviates from that of a circle by only 0.19% and therefore may be

ignored): According to Fig. 23, the minimum cross-sectional increase for an 8.4% diametral decrease is about 2.5%, and therefore the actual O-ring width when installed in the groove will be

$$W'_{min} = 1.025(0.070 - 0.003) = 0.0687 \text{ in.}$$
$$W'_{max} = 1.025(0.070 + 0.003) = 0.0748 \text{ in.}$$

7. w, groove width:

$$w_{min} = W'_{max}(1.05)$$

$$= 0.0748(1.05) = 0.0785 \text{ in.}$$

$w = 0.079 \pm 0.001$ in. is satisfactory

8. Actual squeeze of O-ring:

a.

$$\text{Minimum} = W'_{min} - \frac{G_{max} - S_{min}}{2}$$

$$= 0.0687 - \frac{0.693 - 0.562}{2}$$

$$= 0.0032 \text{ in.}$$

b.

$$\text{Maximum} = W'_{max} - \frac{G_{min} - S_{max}}{2}$$

$$= 0.0748 - \frac{0.691 - 0.563}{2}$$

$$= 0.0108 \text{ in.}$$

9. Average reciprocating speed of O-ring on shaft:

Reciprocating speed $= 2S(\tan\theta)N$

$$= (2) \frac{(0.563 \text{ in.})(\tan 5°)}{12 \text{ in./ft}} \left(3000 \frac{\text{rev}}{\text{min}}\right)$$

$$= 24.6 \text{ fpm}$$

This speed is acceptable because it is greater than 20 fpm.

10. Frictional torque:

a. Friction due to differential pressure: Referring to Fig. 17, where the O-ring mean diameter is

$$\frac{0.614 + 0.754}{2} = 0.684 \text{ in. diam.}$$

and the differential pressure is 500 psi, we find that

$$F_H = 3.5 \text{ lb}$$

    b. Friction due to cross-sectional squeeze: Referring to Fig. 18, where the O-ring ID is 0.5625 diameter, and the maximum percent cross-sectional squeeze is

$$\frac{0.0108}{0.0748} \times 100\% = 14.4\%$$

we find that

$$F_C = 1.5 \text{ lb}$$

    c. Total friction:

$$F = 3.5 + 1.5 = 5.0 \text{ lb}$$

    d. Frictional torque:

$$T = 5.0 \left(\frac{0.5625}{2} \right) = 1.4 \text{ in.-lb}$$

11. PV parameter:

$$PV = (\Delta P)(\pi S N)$$

$$= (500 \text{ psi}) \left(\pi \, \frac{0.563 \text{ in.}}{12 \text{ in./ft}} \right) \left(3000 \, \frac{\text{rev}}{\text{min}} \right)$$

$$= 221,000 \text{ psi-fpm}$$

According to Fig. 26, there O-ring seals should function indefinitely.

A slanted-groove design that involved a shaft twice the diameter used in Design Example 12, but rotating at half the speed, incorporated a slant angle of 2.5 degrees instead of the 5 degrees appearing in Fig. 29. The surface speed of this shaft, N, was then 441.8 fpm, the same as for the shaft of Design Example 12, but the reciprocating speed of the O-ring across the shaft surface was now only half, 12.3 fpm, this being

$$2(1.125)(\tan 2.5°) \, \frac{\text{in.}}{\text{rev}} \, \frac{\text{ft}}{12 \text{ in.}} \left(1500 \, \frac{\text{rev}}{\text{min}} \right) = 12.3 \text{ fpm}$$

This design resulted in the same equivalent reduction in shaft wear as compared to a vertical groove design, but without any apparent reduction in torque. Evidently, slanting the O-ring by 2.5 degrees was enough to adequately distribute the wear load of

the O-ring and extend the life of the shaft, but not enough
to effectively reduce the running friction at the O-ring seal
interface. The reciprocation speed of the O-ring was probably
slow enough to produce a "striction" type of friction caused by
the constant sweeping of fresh lubrication across the seal area.
The "striction" or static friction is caused when the O-ring ac-
celerates too slowly from its end-stroke position. In Design Example 12 where the slant angle was such as to produce a faster
O-ring reciprocation speed (24.6 fpm), this "striction" torque
could not occur and the net result was a reduction in torque required to rotate the shaft.

Therefore, it is hypothesized that in addition to reducing
shaft wear, a slanted O-ring groove may be designed to reduce
shaft torque only if the proper slant angle is chosen. This slant
angle should be such that the reciprocation speed across the
shaft surface is greater than 20 fpm. It is theorized that the
minimum reciprocation speed is required so that the "striction"
effects of the stop-and-go action of the O-ring, relative to the
axis of the shaft, do not significantly add to the frictional torque
required to rotate the shaft.

F. General Conclusions—Seal Life

The use of peripheral compression and self-lubrication have vastly
extended the applicability of elastomeric O-rings as rotary seals.
O-rings in peripheral compression have successfully been used to
prevent the Gow-Joule failure phenomenon, while the additional
design feature of slanting the O-ring has extended the environmental limits of rotational speeds, pressures, and seal life.

Rotary O-ring glands should always be designed to incorporate
peripheral compression. Such gland designs basically reduce the
standard groove diameter of the O-ring housing by 8% to keep the
O-ring in compression. O-ring failure will not occur due to the
Gow-Joule effect if the net forces on the O-ring result in peripheral compression. Designing all rotary O-ring seals for peripheral compression increases the safety factor in which the compressive stresses are always greater than the tensile stresses induced by the differential pressure and the heat generated by
shaft rotation.

Figure 26 predicts the service life of O-rings used in peripheral compression for perpendicular glands. Figure 30 presents
the service life for slanted O-ring glands. Notice that the service
life for slanted O-ring glands is approximately four times the
service life for O-rings in perpendicular glands. This is because

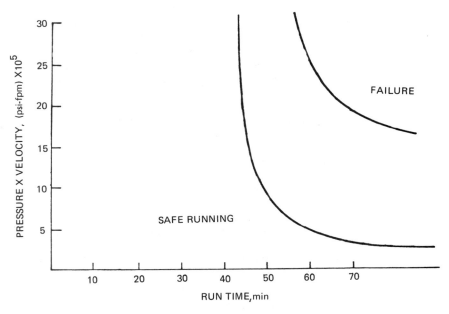

Figure 30.   Extreme PV values versus safe running time—slanted groove

the slanted O-rings sweep lubricant across the sealing surface of the shaft, reducing the dynamic friction (Fig. 28). There have been cases in which slanted O-ring glands have actually extended service life at PV values of 2.2×10^5 psi-fpm to 15 times that of perpendicular glands. This was accomplished by using 70° Shore hardness Buna-N O-rings at these lower PV values instead of the normally used 80° Shore hardness Buna-N graphite-impregnated O-rings required for the higher PV values, 12×10^5 psi-fpm.

In general, slanting the O-ring gland extends seal life by four times that of a perpendicular O-ring gland. The only disadvantage of the slanted, peripheral compression gland is its initial manufacturing cost, but once production tooling is made, cost becomes comparable to the perpendicular, peripheral compression gland.

II. PRACTICAL DESIGN DATA

A. Actual Rotary Seal Hardware

Figure 31 shows a typical rotary O-ring gland designed for peripheral compression and oil lubrication. The gland housing is designed to "float" about the shaft and is made of SAE 660 bronze to dissipate the heat away from the oil and O-ring seals. The shaft is actually a stainless steel (AISI 303) collar that is chrome plated at the O-ring interface. The chrome plate is approximately 0.005 in. thick and polished to a surface finish of 16 rms. Chrome plate provides the O-rings with a hard, smooth surface which has the optimum microscopic porosity to hold the oil that minimizes the friction and dissipates the heat.

The rotary O-ring gland design presented in Fig. 31 has been used in a production undersea vehicle for the last 16 years. The tough seawater environment has proven this design to be very reliable in providing a positive seal at extended O-ring life cycles.

Figure 31. Chrome shaft with bronze seal housing (Courtesy of U.S. Naval Ocean Systems Center, San Diego, Calif.)

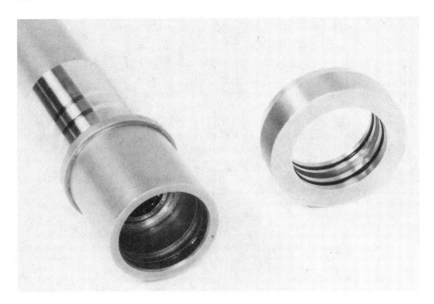

Figure 32. Chrome shaft with slanted seal housing (Courtesy of
U.S. Naval Ocean Systems Center, San Diego, Calif.)

Figure 32 shows the actual seal hardware of Design Example
12. The shaft is chrome plated at the O-ring interface and
shows two wide, polished witness marks corresponding to the
slanted O-rings of the seal. The seal housing is made of SAE 660
bronze to dissipate heat. A needle bearing supports the shaft
on the chrome surface directly behind the seal housing. The
designer must always provide a bearing or some support mechanism
for the shaft. The seal housing must never be used to support
the shaft, even if the O-ring grooves are machined into the base
housing. Figure 33 shows the seal housing and needle bearing
installed in the base housing. Dimensions are specified such that
the seal housing may "float" about the shaft (not shown). The
seal housing can move radially 0.0035 ± 0.0010 in. and axially
0.011 ± 0.009 in. The radial clearance must be large enough to
compensate for assembly eccentricities caused by tolerance stack-
up between the needle bearing and base housing and the seal
housing and base housing. Sufficient radial clearance will ensure
proper alignment (floating) of the seal housing without damage
to the rotating shaft.

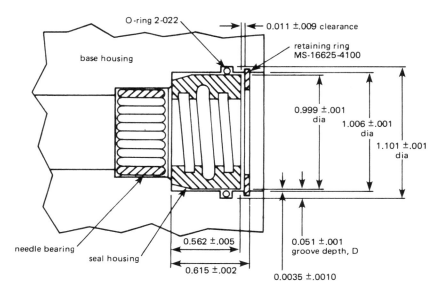

Figure 33. Dimensions for floating seal (shaft not installed).

B. O-Ring Material for Rotary Seal Applications

Various elastomeric materials have been tried in rotary seal applications to extend the seal life of O-rings. Materials tried have included fluorocarbon (Parker No. V747-75), neoprene (Parker No. C557-70), nitrile (Buna-N, Parker No. N532-80), polyurethane (Parker No. P642-70), and graphite-impregnated nitrile (Buna-N, Parker No. N256-85). Tests have shown that nitrile N532-80 lasts as long as fluorocarbon V747-75 and polyurethane P642-70, but neoprene C557-70 lasts only one-fourth of the time before leakage and catastrophic failure of the O-rings occurs. The graphite-impregnated nitrile O-rings last at least twice as long as the neoprene, nitrile, or polyurethane O-rings, but the graphite-impregnated nitrile O-rings tend to wear the shaft at a greater rate.

The order of best physical resistance (Chart 1D, Chap. 2) predicts nitrile as having the best physical resistance, while also being compatible with oil, the oil being required for the dual-O-ring and oil annulus seal design for rotary seal applications. The designer should realize that the O-ring material must be compatible with the fluid being sealed. Such charts as 1D can help in the

selection of the proper elastomeric material, but the various potential materials should actually be tested in order to optimize
seal life and determine the type of material failure. Type of
failure may become very important when abrupt failure cannot be
tolerated. If material deterioration is slow enough to indicate
partial leakage, the O-rings can be replaced before catastrophic
leakage occurs. Testing potential materials will indicate the type
of failure to be expected. Failures due to extrusion and heat in-
compatibility usually occur quickly, while failures due to wear
and general fatigue tend to progress more slowly.

C. Assembly Procedure for Dual Peripherally Compressed O-Ring, Oil Annulus, Rotary Seal Design

The following procedure describes how to assemble the rotary
shaft seal of Fig. 32 and can generally be applied to any rotary
seal design that incorporates O-rings in peripheral compression
and/or an oil annulus:

1. The O-rings chosen to induce peripheral compression will
appear to be too large circumferentially for the grooves in the
seal housing but with a little effort they can be worked into
place. First, inspect the two O-ring grooves in the housing to
make sure that they are clean. Lightly coat the inside of the
two grooves with a light lubricant such as Wynn oil 8P-27 (70
parts by volume 1245 neutral oil and 30 parts by volume 594 high-
performance lube supplement; Wynn Oil Company, Azusa, CA
91702) to facilitate installation of the O-rings. Warm the O-rings
to 135°F prior to installation. To install an O-ring in its groove,
press the O-ring between the thumb and forefinger to form an
ellipse. Push the "end" of the ellipse into the O-ring groove, and
with the forefinger and the thumb, work the O-ring radially into
the groove. A small loop in the seal which does not slide into
the groove remains. Push this loop gently into the groove with
one finger and with a blunt, very smooth probe force the O-ring
material from one side of the loop, at the point where it emerges
from the groove, to the other side with a motion that tends to
distribute the O-ring material around the groove. The O-ring
material flows slowly. Nevertheless, with steady pressure, the
O-ring will finally "flow" into place. After all of the O-ring is
seated in the groove, again use the finger and thumb to make
sure that the O-ring material is distributed, without twisting,
around the circumference of the groove.

Figure 34. Oil filling of rotary seal

2. Coat the outer shaft sealing surface with a light coat of Wynn oil 8P-27 or equivalent. Slip the shaft seal housing onto the shaft to engage the first O-ring.

3. Holding the shaft in a vertical position, the annulus between the two O-rings in the seal housing should be filled to 50 to 60% capacity with Wynn oil 8P-27 or equivalent. This is accomplished by using a calibrated hypodermic syringe with a blunt needle (Fig. 34).

4. After the oil is injected into the annulus, the seal housing is slid farther onto the shaft such that the oil is retained by both O-rings.

D. Peripheral Compression Rotary Seal Design Tables

Design Tables 21 and 22 give dimensions for rotary shaft seal designs incorporating O-rings in peripheral compression for most current-size O-rings. The rotary shaft seal design is depicted in Fig. 35 and defines the shaft, groove, and housing diameters presented in Table 22 and the depth and width presented in Table 21. Gland dimensions are such that the peripheral compression incurred by the O-ring averages 7% and the squeeze on the O-ring cross section ranges from 6 to 16%. The peripheral compression and cross-sectional squeeze for each O-ring size are given in Table 23.

Design Tables 21 and 22 reflect the theoretical criteria presented for both perpendicular and slanted O-ring glands, except that the O-ring cross-sectional squeeze is slightly greater than

Table 21.   Selection of O-Ring Cross Section Based on Shaft
Speed

O-ring cross-sectional diameter (in.)	Suggested shaft speed (fpm)	Nominal gland depth (in.)	O-ring groove width (in.)
0.070	Less than 1000	0.064	0.079
0.103	Less than 600	0.095	0.115
0.139	Less than 400	0.129	0.155
0.210	Less than 200	0.193	0.233
0.275	Less than 100	0.253	0.304

that calculated in Design Examples 11 and 12. The design tables
are based on the most commonly used O-ring sizes and include the
currently increased tolerances on the inside diameter of the O-ring
per standard AS568A. The larger tolerances for the inside diameter of the O-rings accommodate a greater range of shrinkage
rates for the more recent elastomeric materials. These tolerances
are greater than those presented in Table 13, and should generally be used unless the actual diamensions of the O-ring are
known.

Table 21 should be consulted for selecting the O-ring cross-
sectional diameter. In most cases, the smallest cross-sectional
diameter should be chosen to minimize the amount of elastic material contacting the shaft and therefore, minimize the amount of
heat generated at the O-ring/shaft interface. In Design Example
13, the shaft speed is 485 fpm. According to Table 21, either
the 0.070-in. or 0.103-in.-cross-section O-ring can be used. The
smaller-cross-section O-ring is selected, but if the problem statement had required a shaft larger than that listed for the 0.070-in.-
cross-section O-ring (4.854 in.), the 0.103-in.-cross-section
O-rings could be used (2-160 to 2-178 O-rings), depending on the
shaft size and shaft speed.

Tables 21 and 22 specify the dimensions depicted in Fig. 35.
The dimensions are such that the diametral clearance between the
shaft and the seal housing bore is constant throughout the tables,
0.006 in., to prevent O-ring extrusion at 1200 psi. The diametral
clearance between the seal housing and the base should be slightly greater than that between the shaft and the housing, to allow
maximum flotation of the seal housing about the shaft. A diametral clearance of 0.010 in. between the seal housing and the base
is adequate. The use of the design tables is not recommended for

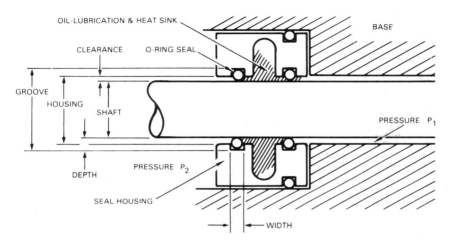

Figure 35. Rotary O-ring gland of peripheral compression and oil annulus

pressures greater than 1200 psi or for shaft speeds greater than 1000 fpm. When pressures are greater than this, the designer may modify the dimensions specified in Table 22 to reduce the housing bore, but the designer should realize that in so doing the concentricity between the shaft and the seal housing becomes more of a factor. The designer may want to slant the O-ring grooves as shown in Fig. 27 if shaft speeds are greater than 1000 fpm. Slanting the O-ring grooves, as specified in Sec. I.E, does not significantly effect the shaft and O-ring groove dimensions specified in Table 22.

Design Example 13 presents the use of Tables 21 to 23. The size of the O-ring is chosen according to the shaft speed and size. Shaft, groove, and housing diameters are selected from the design Table 22 and the peripheral compression and cross-sectional squeeze incurrend by the O-ring are presented in Table 23 for the designer's information. The designer must specify surface condition and concentricity together with dimensions. Generally speaking, shafts should be ground to a finish of 16 rms, while the surfaces of the O-ring grooves should be 32 to 63

Table 22. Rotary Shaft Seal Design Table (Floating Housing)
(Reference Fig. 35)

MAXIMUM CAPABILITY: 1200 psi, 1000 rpm

Unless specified all dimensions are in inches
[ ] denotes tolerances,+ or -

W (cross-sectional dia.) = .070[.003]
██

O-RING	ID	OD	SHAFT DIA [.0005]	GROOVE DIA [.001]	HOUSING BORE [.0005]
2-010	0.239[.005]	0.379	0.224	0.351	0.230
2-011	0.301[.005]	0.441	0.281	0.408	0.287
2-012	0.364[.005]	0.504	0.340	0.467	0.346
2-013	0.426[.005]	0.566	0.397	0.524	0.403
2-014	0.489[.005]	0.629	0.455	0.582	0.461
2-015	0.551[.007]	0.691	0.513	0.640	0.519
2-016	0.614[.009]	0.754	0.571	0.698	0.577
2-017	0.676[.009]	0.816	0.629	0.756	0.635
2-018	0.739[.009]	0.879	0.687	0.814	0.693
2-019	0.801[.009]	0.941	0.744	0.871	0.750
2-020	0.864[.009]	1.004	0.803	0.930	0.809
2-021	0.926[.009]	1.066	0.860	0.987	0.866
2-022	0.989[.010]	1.129	0.918	1.045	0.924
2-023	1.051[.010]	1.191	0.976	1.103	0.982
2-024	1.114[.010]	1.254	1.034	1.161	1.040
2-025	1.176[.011]	1.316	1.092	1.219	1.098
2-026	1.239[.011]	1.379	1.150	1.277	1.156
2-027	1.301[.011]	1.441	1.207	1.334	1.213
2-028	1.364[.013]	1.504	1.266	1.393	1.272
2-029	1.489[.013]	1.629	1.381	1.508	1.387
2-030	1.614[.013]	1.754	1.497	1.624	1.503
2-031	1.739[.015]	1.879	1.613	1.740	1.619
2-032	1.864[.015]	2.004	1.729	1.856	1.735
2-033	1.989[.018]	2.129	1.844	1.971	1.850
2-034	2.114[.018]	2.254	1.960	2.087	1.966
2-035	2.239[.018]	2.379	2.076	2.203	2.082
2-036	2.364[.018]	2.504	2.192	2.319	2.198
2-037	2.489[.018]	2.629	2.307	2.434	2.313
2-038	2.614[.020]	2.754	2.423	2.550	2.429
2-039	2.739[.020]	2.879	2.539	2.666	2.545
2-040	2.864[.020]	3.004	2.654	2.781	2.660
2-041	2.989[.024]	3.129	2.770	2.897	2.776
2-042	3.239[.024]	3.379	3.002	3.129	3.008
2-043	3.489[.024]	3.629	3.233	3.360	3.239
2-044	3.739[.027]	3.879	3.465	3.592	3.471
2-045	3.989[.027]	4.129	3.696	3.823	3.702
2-046	4.239[.030]	4.379	3.928	4.055	3.934
2-047	4.489[.030]	4.629	4.159	4.286	4.165
2-048	4.739[.030]	4.879	4.391	4.518	4.397
2-049	4.989[.037]	5.129	4.622	4.749	4.628
2-050	5.239[.037]	5.379	4.854	4.981	4.860

Table 22.   (Continued)

MAXIMUM CAPABILITY: 1200 psi, 600 fpm

Unless specified all dimensions are in inches
[ ] denotes tolerances,+ or -

W (cross-sectional dia.) = .103[.003]

O-RING	ID	OD	SHAFT DIA [.0005]	GROOVE DIA [.001]	HOUSING BORE [.0005]
2-110	0.362[.005]	0.568	0.336	0.526	0.342
2-111	0.424[.005]	0.630	0.394	0.583	0.400
2-112	0.487[.005]	0.693	0.452	0.642	0.458
2-113	0.549[.007]	0.755	0.509	0.699	0.515
2-114	0.612[.009]	0.818	0.568	0.757	0.574
2-115	0.674[.009]	0.880	0.625	0.815	0.631
2-116	0.737[.009]	0.943	0.683	0.873	0.689
2-117	0.799[.010]	1.005	0.741	0.931	0.747
2-118	0.862[.010]	1.068	0.799	0.989	0.805
2-119	0.924[.010]	1.130	0.857	1.046	0.863
2-120	0.987[.010]	1.193	0.915	1.105	0.921
2-121	1.049[.010]	1.255	0.972	1.162	0.978
2-122	1.112[.010]	1.318	1.031	1.220	1.037
2-123	1.174[.012]	1.380	1.088	1.278	1.094
2-124	1.237[.012]	1.443	1.146	1.336	1.152
2-125	1.299[.012]	1.505	1.204	1.394	1.210
2-126	1.362[.012]	1.568	1.262	1.452	1.268
2-127	1.424[.012]	1.630	1.320	1.509	1.326
2-128	1.487[.012]	1.693	1.378	1.568	1.384
2-129	1.549[.015]	1.755	1.435	1.625	1.441
2-130	1.612[.015]	1.818	1.494	1.683	1.500
2-131	1.674[.015]	1.880	1.551	1.741	1.557
2-132	1.737[.015]	1.943	1.609	1.799	1.615
2-133	1.799[.015]	2.005	1.667	1.856	1.673
2-134	1.862[.015]	2.068	1.725	1.915	1.731
2-135	1.925[.017]	2.131	1.783	1.973	1.789
2-136	1.987[.017]	2.193	1.841	2.031	1.847
2-137	2.050[.017]	2.256	1.899	2.089	1.905
2-138	2.112[.017]	2.318	1.957	2.146	1.963
2-139	2.175[.017]	2.381	2.015	2.205	2.021
2-140	2.237[.017]	2.443	2.072	2.262	2.078
2-141	2.300[.020]	2.506	2.131	2.320	2.137
2-142	2.362[.020]	2.568	2.188	2.378	2.194
2-143	2.425[.020]	2.631	2.246	2.436	2.252
2-144	2.487[.020]	2.693	2.304	2.494	2.310
2-145	2.550[.020]	2.756	2.362	2.552	2.368
2-146	2.612[.020]	2.818	2.420	2.609	2.426
2-147	2.675[.022]	2.881	2.478	2.668	2.484
2-148	2.737[.022]	2.943	2.535	2.725	2.541
2-149	2.800[.022]	3.006	2.594	2.783	2.600
2-150	2.862[.022]	3.068	2.651	2.841	2.657
2-151	2.987[.024]	3.193	2.767	2.956	2.773
2-152	3.237[.024]	3.443	2.998	3.188	3.004

Table 22. (Continued)

MAXIMUM CAPABILITY: 1200 psi, 600 fpm

Unless specified all dimensions are in inches
[ ] denotes tolerances,+ or -

W (cross-sectional dia.) = .103[.003]

O-RING	ID	OD	SHAFT DIA [.0005]	GROOVE DIA [.001]	HOUSING BORE [.0005]
2-153	3.487[.024]	3.693	3.230	3.419	3.236
2-154	3.737[.028]	3.943	3.461	3.651	3.467
2-155	3.987[.028]	4.193	3.693	3.882	3.699
2-156	4.237[.030]	4.443	3.924	4.114	3.930
2-157	4.487[.030]	4.693	4.156	4.345	4.162
2-158	4.737[.030]	4.943	4.387	4.577	4.393
2-159	4.987[.035]	5.193	4.619	4.808	4.625
2-160	5.237[.035]	5.443	4.850	5.040	4.856
2-161	5.487[.035]	5.693	5.082	5.271	5.088
2-162	5.737[.035]	5.943	5.313	5.503	5.319
2-163	5.987[.035]	6.193	5.45	5.734	5.551
2-164	6.237[.040]	6.443	5.776	5.966	5.782
2-165	6.487[.040]	6.693	6.008	6.197	6.014
2-166	6.737[.040]	6.943	6.239	6.429	6.245
2-167	6.987[.040]	7.193	6.470	6.660	6.476
2-168	7.237[.045]	7.443	6.702	6.892	6.708
2-169	7.487[.045]	7.693	6.933	7.123	6.939
2-170	7.737[.045]	7.943	7.165	7.355	7.171
2-171	7.987[.045]	8.193	7.396	7.586	7.402
2-172	8.237[.050]	8.443	7.628	7.818	7.634
2-173	8.487[.050]	8.693	7.859	8.049	7.865
2-174	8.737[.050]	8.943	8.091	8.281	8.097
2-175	8.987[.050]	9.193	8.322	8.512	8.328
2-176	9.237[.055]	9.443	8.554	8.744	8.560
2-177	9.487[.055]	9.693	8.785	8.975	8.791
2-178	9.737[.055]	9.943	9.017	9.206	9.023

MAXIMUM CAPABILITY: 1200 psi, 400 fpm

Unless specified all dimensions are in inches
[ ] denotes tolerances,+ or -

W (cross-sectional dia.) = .139[.004]

O-RING	ID	OD	SHAFT DIA [.0005]	GROOVE DIA [.001]	HOUSING BORE [.0005]
2-210	0.734[.010]	1.012	0.679	0.937	0.685
2-211	0.796[.010]	1.074	0.736	0.994	0.742
2-212	0.859[.010]	1.137	0.795	1.053	0.801
2-213	0.921[.010]	1.199	0.852	1.110	0.858
2-214	0.984[.010]	1.262	0.910	1.169	0.916

Table 22.   (Continued)

MAXIMUM CAPABILITY: 1200 ps1, 400 fpm

Unless specified all dimensions are in inches
[ ] denotes tolerances,+ or -

W (cross-sectional dia.) = .139[.004]
■■■■■■■■■■■■■■■■■■■■■■■■■■■■■■■■■■■■■■

O-RING	ID	OD	SHAFT DIA [.0005]	GROOVE DIA [.001]	HOUSING BORE [.0005]
2-215	1.046[.010]	1.324	0.968	1.226	0.974
2-216	1.109[.012]	1.387	1.026	1.284	1.032
2-217	1.171[.012]	1.449	1.084	1.342	1.090
2-218	1.234[.012]	1.512	1.142	1.400	1.148
2-219	1.296[.012]	1.574	1.199	1.457	1.205
2-220	1.359[.012]	1.637	1.258	1.516	1.264
2-221	1.421[.012]	1.699	1.315	1.573	1.321
2-222	1.484[.015]	1.762	1.373	1.631	1.379
2-223	1.609[.015]	1.887	1.489	1.747	1.495
2-224	1.734[.015]	2.012	1.605	1.863	1.611
2-225	1.859[.018]	2.137	1.721	1.979	1.727
2-226	1.984[.018]	2.262	1.836	2.094	1.842
2-227	2.109[.018]	2.387	1.952	2.210	1.958
2-228	2.234[.020]	2.512	2.068	2.326	2.074
2-229	2.359[.020]	2.637	2.184	2.442	2.190
2-230	2.484[.020]	2.762	2.299	2.557	2.305
2-231	2.609[.020]	2.887	2.415	2.673	2.421
2-232	2.734[.024]	3.012	2.531	2.789	2.537
2-233	2.859[.024]	3.137	2.647	2.905	2.653
2-234	2.984[.024]	3.262	2.762	3.020	2.768
2-235	3.109[.024]	3.387	2.878	3.136	2.884
2-236	3.234[.024]	3.512	2.994	3.252	3.000
2-237	3.359[.024]	3.637	3.109	3.368	3.115
2-238	3.484[.024]	3.762	3.225	3.483	3.231
2-239	3.609[.028]	3.887	3.341	3.599	3.347
2-240	3.734[.028]	4.012	3.457	3.715	3.463
2-241	3.859[.028]	4.137	3.572	3.831	3.578
2-242	3.984[.028]	4.262	3.688	3.946	3.694
2-243	4.109[.028]	4.387	3.804	4.062	3.810
2-244	4.234[.030]	4.512	3.920	4.178	3.926
2-245	4.359[.030]	4.637	4.035	4.294	4.041
2-246	4.484[.030]	4.762	4.151	4.409	4.157
2-247	4.609[.030]	4.887	4.267	4.525	4.273
2-248	4.734[.030]	5.012	4.383	4.641	4.389
2-249	4.859[.035]	5.137	4.498	4.756	4.504
2-250	4.894[.035]	5.172	4.531	4.789	4.537
2-251	5.109[.035]	5.387	4.730	4.988	4.736
2-252	5.234[.035]	5.512	4.846	5.104	4.852
2-253	5.359[.035]	5.637	4.961	5.219	4.967
2-254	5.484[.035]	5.762	5.077	5.335	5.083
2-255	5.609[.035]	5.887	5.193	5.451	5.199
2-256	5.734[.035]	6.012	5.309	5.567	5.315
2-257	5.859[.035]	6.137	5.424	5.682	5.430
2-258	5.984[.035]	6.262	5.540	5.798	5.546
2-259	6.234[.040]	6.512	5.772	6.030	5.778

Table 22.   (Continued)

MAXIMUM CAPABILITY: 1200 psi, 400 fpm

Unless specified all dimensions are in inches
[ ] denotes tolerances,+ or -

W (cross-sectional dia.) = .139[.004]
■■

O-RING	ID	OD	SHAFT DIA [.0005]	GROOVE DIA [.001]	HOUSING BORE [.0005]
2-260	6.484[.040]	6.762	6.003	6.261	6.009
2-261	6.734[.040]	7.012	6.234	6.493	6.240
2-262	6.984[.040]	7.262	6.466	6.724	6.472
2-263	7.234[.045]	7.512	6.697	6.956	6.703
2-264	7.484[.045]	7.762	6.929	7.187	6.935
2-265	7.734[.045]	8.012	7.160	7.419	7.166
2-266	7.984[.045]	8.262	7.392	7.650	7.398
2-267	8.234[.050]	8.512	7.623	7.881	7.629
2-268	8.484[.050]	8.762	7.855	8.113	7.861
2-269	8.734[.050]	9.012	8.086	8.344	8.092
2-270	8.984[.050]	9.262	8.318	8.576	8.324
2-271	9.234[.055]	9.512	8.549	8.807	8.555
2-272	9.484[.055]	9.762	8.781	9.039	8.787
2-273	9.734[.055]	10.012	9.012	9.270	9.018
2-274	9.984[.055]	10.262	9.244	9.502	9.250

MAXIMUM CAPABILITY: 1200 psi, 200 fpm

Unless specified all dimensions are in inches
[ ] denotes tolerances,+ or -

W (cross-sectional dia.) = .210[.005]
■■■■■■■■■■■■■■■■■■■■■■■■■■■■■■■■■■■■■■■

O-RING	ID	OD	SHAFT DIA [.0005]	GROOVE DIA [.001]	HOUSING BORE [.0005]
2-315	0.787[.010]	1.207	0.725	1.118	0.731
2-316	0.850[.010]	1.270	0.783	1.176	0.789
2-317	0.912[.010]	1.332	0.840	1.233	0.846
2-318	0.975[.010]	1.395	0.899	1.292	0.905
2-319	1.037[.010]	1.457	0.956	1.349	0.962
2-320	1.100[.012]	1.520	1.014	1.407	1.020
2-321	1.162[.012]	1.582	1.072	1.465	1.078
2-322	1.225[.012]	1.645	1.130	1.523	1.136
2-323	1.287[.012]	1.707	1.188	1.581	1.194
2-324	1.350[.012]	1.770	1.246	1.639	1.252
2-325	1.475[.015]	1.895	1.362	1.755	1.368
2-326	1.600[.015]	2.020	1.477	1.870	1.483
2-327	1.725[.015]	2.145	1.593	1.986	1.599
2-328	1.850[.015]	2.270	1.709	2.102	1.715
2-329	1.975[.018]	2.395	1.825	2.218	1.831
2-330	2.100[.018]	2.520	1.940	2.333	1.946
2-331	2.225[.018]	2.645	2.056	2.449	2.062
2-332	2.350[.018]	2.770	2.172	2.565	2.178

Table 22.   (Continued)

MAXIMUM CAPABILITY: 1200 psi, 200 fpm

Unless specified all dimensions are in inches
[ ] denotes tolerances,+ or -

W (cross-sectional dia.) = .210[.005]
▓███
O-RING	ID	OD	SHAFT DIA [.0005]	GROOVE DIA [.001]	HOUSING BORE [.0005]
2-333	2.475[.020]	2.895	2.288	2.681	2.294
2-334	2.600[.020]	3.020	2.403	2.796	2.409
2-335	2.725[.020]	3.145	2.519	2.912	2.525
2-336	2.850[.020]	3.270	2.635	3.028	2.641
2-337	2.975[.024]	3.395	2.751	3.144	2.757
2-338	3.100[.024]	3.520	2.866	3.259	2.872
2-339	3.225[.024]	3.645	2.982	3.375	2.988
2-340	3.350[.024]	3.770	3.098	3.491	3.104
2-341	3.475[.024]	3.895	3.213	3.606	3.219
2-342	3.600[.028]	4.020	3.329	3.722	3.335
2-343	3.725[.028]	4.145	3.445	3.838	3.451
2-344	3.850[.028]	4.270	3.561	3.954	3.567
2-345	3.975[.028]	4.395	3.676	4.069	3.682
2-346	4.100[.028]	4.520	3.792	4.185	3.798
2-347	4.225[.030]	4.645	3.908	4.301	3.914
2-348	4.350[.030]	4.770	4.024	4.417	4.030
2-349	4.475[.030]	4.895	4.139	4.532	4.145
2-350	4.600[.030]	5.020	4.255	4.648	4.261
2-351	4.725[.030]	5.145	4.371	4.764	4.377
2-352	4.850[.030]	5.270	4.487	4.880	4.493
2-353	4.975[.037]	5.395	4.602	4.995	4.608
2-354	5.100[.037]	5.520	4.718	5.111	4.724
2-355	5.225[.037]	5.645	4.834	5.227	4.840
2-356	5.350[.037]	5.770	4.950	5.343	4.956
2-357	5.475[.037]	5.895	5.065	5.458	5.071
2-358	5.600[.037]	6.020	5.181	5.574	5.187
2-359	5.725[.037]	6.145	5.297	5.690	5.303
2-360	5.850[.037]	6.270	5.413	5.806	5.419
2-361	5.975[.037]	6.395	5.528	5.921	5.534
2-362	6.225[.040]	6.645	5.760	6.153	5.766
2-363	6.475[.040]	6.895	5.991	6.384	5.997
2-364	6.725[.040]	7.145	6.223	6.616	6.229
2-365	6.975[.040]	7.395	6.454	6.847	6.460
2-366	7.225[.045]	7.645	6.686	7.079	6.692
2-367	7.475[.045]	7.895	6.917	7.310	6.923
2-368	7.725[.045]	8.145	7.149	7.542	7.155
2-369	7.975[.045]	8.395	7.380	7.773	7.386
2-370	8.225[.050]	8.645	7.612	8.005	7.618
2-371	8.475[.050]	8.895	7.843	8.236	7.849
2-372	8.725[.050]	9.145	8.075	8.468	8.081
2-373	8.975[.050]	9.395	8.306	8.699	8.312
2-374	9.225[.055]	9.645	8.538	8.931	8.544
2-375	9.475[.055]	9.895	8.769	9.162	8.775
2-376	9.725[.055]	10.145	9.001	9.394	9.007
2-377	9.975[.055]	10.395	9.232	9.625	9.238

Table 22.   (Continued)

MAXIMUM CAPABILITY: 1200 psi, 100 fpm

Unless specified all dimensions are in inches
[ ] denotes tolerances,+ or -

W (cross-sectional dia.) = .275[.006]
==

O-RING	ID	OD	SHAFT DIA [.0005]	GROOVE DIA [.001]	HOUSING BORE [.0005]
2-425	4.475[.033]	5.025	4.136	4.653	4.142
2-426	4.600[.033]	5.150	4.252	4.769	4.258
2-427	4.725[.033]	5.275	4.368	4.884	4.374
2-428	4.850[.033]	5.400	4.484	5.000	4.490
2-429	4.975[.037]	5.525	4.599	5.116	4.605
2-430	5.100[.037]	5.650	4.715	5.231	4.721
2-431	5.225[.037]	5.775	4.831	5.347	4.837
2-432	5.350[.037]	5.900	4.946	5.463	4.952
2-433	5.475[.037]	6.025	5.062	5.579	5.068
2-434	5.600[.037]	6.150	5.178	5.694	5.184
2-435	5.725[.037]	6.275	5.294	5.810	5.300
2-436	5.850[.037]	6.400	5.409	5.926	5.415
2-437	5.975[.037]	6.525	5.525	6.042	5.531
2-438	6.225[.040]	6.775	5.757	6.273	5.763
2-439	6.475[.040]	7.025	5.988	6.505	5.994
2-440	6.725[.040]	7.275	6.220	6.736	6.226
2-441	6.975[.040]	7.525	6.451	6.968	6.457
2-442	7.225[.045]	7.775	6.683	7.199	6.689
2-443	7.475[.045]	8.025	6.914	7.431	6.920
2-444	7.725[.045]	8.275	7.146	7.662	7.152
2-445	7.975[.045]	8.525	7.377	7.894	7.383
2-446	8.475[.055]	9.025	7.840	8.356	7.846
2-447	8.975[.055]	9.525	8.303	8.819	8.309
2-448	9.475[.055]	10.025	8.766	9.282	8.772
2-449	9.975[.055]	10.525	9.229	9.745	9.235

rms. Concentricities must be held to 50% of the minimum O-ring
cross-sectional squeeze (i.e., 0.002 in. for 0.070-in.-cross-section
O-rings to 0.006 for 0.210-in.-cross-section O-rings. In order to
maintain shaft concentricity, the shaft support mechanism (bearing) must be located as close to the seal housing as possible. The
designer should realize that the seal housing must float about the
rotating shaft and not support the shaft. O-rings cannot be used
to support a rotating shaft. This must be accomplished by another,
independent mechanism, such as journal or rolling-element bearings. The better the shaft is supported and the straighter the
shaft is, the better will be the running concentricity of the shaft
and the longer will be the life of the O-ring seals.

Table 23. Rotary Shaft Seal Design Table (O-Ring Squeeze and Peripheral Compression) (Reference Fig. 35)

W (cross-sectional dia.) = .070[.003]

SQUEEZE, (%)		PERIPHERAL COMPRESSION, %		O-RING
MIN	MAX	MIN	MAX	
0.004(6.17)	0.012(15.89)	4.368	10.275	2-010
0.004(6.17)	0.012(15.89)	4.806	9.882	2-011
0.004(6.17)	0.012(15.89)	5.139	9.579	2-012
0.004(6.17)	0.012(15.89)	5.392	9.346	2-013
0.004(6.17)	0.012(15.89)	5.598	9.155	2-014
0.004(6.17)	0.012(15.89)	5.485	9.259	2-015
0.004(6.17)	0.012(15.89)	5.393	9.344	2-016
0.004(6.17)	0.012(15.89)	5.549	9.199	2-017
0.004(6.17)	0.012(15.89)	5.684	9.073	2-018
0.004(6.17)	0.012(15.89)	5.800	8.965	2-019
0.004(6.17)	0.012(15.89)	5.902	8.869	2-020
0.004(6.17)	0.012(15.89)	5.991	8.785	2-021
0.004(6.17)	0.012(15.89)	5.986	8.789	2-022
0.004(6.17)	0.012(15.89)	6.061	8.719	2-023
0.004(6.17)	0.012(15.89)	6.130	8.653	2-024
0.004(6.17)	0.012(15.89)	6.119	8.663	2-025
0.004(6.17)	0.012(15.89)	6.178	8.607	2-026
0.004(6.17)	0.012(15.89)	6.232	8.556	2-027
0.004(6.17)	0.012(15.89)	6.155	8.628	2-028
0.004(6.17)	0.012(15.89)	6.253	8.536	2-029
0.004(6.17)	0.012(15.89)	6.336	8.456	2-030
0.004(6.17)	0.012(15.89)	6.307	8.483	2-031
0.004(6.17)	0.012(15.89)	6.376	8.417	2-032
0.004(6.17)	0.012(15.89)	6.304	8.486	2-033
0.004(6.17)	0.012(15.89)	6.366	8.427	2-034
0.004(6.17)	0.012(15.89)	6.421	8.374	2-035
0.004(6.17)	0.012(15.89)	6.471	8.326	2-036
0.004(6.17)	0.012(15.89)	6.516	8.283	2-037
0.004(6.17)	0.012(15.89)	6.488	8.309	2-038
0.004(6.17)	0.012(15.89)	6.529	8.271	2-039
0.004(6.17)	0.012(15.89)	6.565	8.235	2-040
0.004(6.17)	0.012(15.89)	6.479	8.318	2-241
0.004(6.17)	0.012(15.89)	6.548	8.252	2-042
0.004(6.17)	0.012(15.89)	6.608	8.194	2-043
0.004(6.17)	0.012(15.89)	6.587	8.214	2-044
0.004(6.17)	0.012(15.89)	6.637	8.166	2-045
0.004(6.17)	0.012(15.89)	6.617	8.185	2-046
0.004(6.17)	0.012(15.89)	6.660	8.143	2-047
0.004(6.17)	0.012(15.89)	6.698	8.106	2-048
0.004(6.17)	0.012(15.89)	6.605	8.197	2-049
0.004(6.17)	0.012(15.89)	6.643	8.160	2-050

Table 23.   (Continued)
W (cross-sectional dia.) = .103[.003]
■■

| SQUEEZE,(%) | | PERIPHERAL COMPRESSION,% | | O-RING |
MIN	MAX	MIN	MAX	
0.007(6.46)	0.014(13.14)	5.399	9.339	2-110
0.007(6.46)	0.014(13.14)	5.600	9.152	2-111
0.007(6.46)	0.014(13.14)	5.767	8.996	2-112
0.007(6.46)	0.014(13.14)	5.650	9.105	2-113
0.007(6.46)	0.014(13.14)	5.553	9.195	2-114
0.007(6.46)	0.014(13.14)	5.686	9.071	2-115
0.007(6.46)	0.014(13.14)	5.803	8.962	2-116
0.007(6.46)	0.014(13.14)	5.808	8.956	2-117
0.007(6.46)	0.014(13.14)	5.904	8.866	2-118
0.007(6.46)	0.014(13.14)	5.988	8.787	2-119
0.007(6.46)	0.014(13.14)	6.064	8.716	2-120
0.007(6.46)	0.014(13.14)	6.131	8.652	2-121
0.007(6.46)	0.014(13.14)	6.193	8.593	2-122
0.007(6.46)	0.014(13.14)	6.110	8.671	2-123
0.007(6.46)	0.014(13.14)	6.168	8.617	2-124
0.007(6.46)	0.014(13.14)	6.219	8.567	2-125
0.007(6.46)	0.014(13.14)	6.268	8.521	2-126
0.007(6.46)	0.014(13.14)	6.311	8.479	2-127
0.007(6.46)	0.014(13.14)	6.353	8.440	2-128
0.007(6.46)	0.014(13.14)	6.228	8.559	2-129
0.007(6.46)	0.014(13.14)	6.270	8.519	2-130
0.007(6.46)	0.014(13.14)	6.308	8.483	2-131
0.007(6.46)	0.014(13.14)	6.344	8.448	2-132
0.007(6.46)	0.014(13.14)	6.377	8.417	2-133
0.007(6.46)	0.014(13.14)	6.409	8.386	2-134
0.007(6.46)	0.014(13.14)	6.350	8.443	2-135
0.007(6.46)	0.014(13.14)	6.380	8.414	2-136
0.007(6.46)	0.014(13.14)	6.409	8.386	2-137
0.007(6.46)	0.014(13.14)	6.436	8.360	2-138
0.007(6.46)	0.014(13.14)	6.462	8.335	2-139
0.007(6.46)	0.014(13.14)	6.486	8.312	2-140
0.007(6.46)	0.014(13.14)	6.396	8.398	2-141
0.007(6.46)	0.014(13.14)	6.421	8.374	2-142
0.007(6.46)	0.014(13.14)	6.445	8.351	2-143
0.007(6.46)	0.014(13.14)	6.467	8.330	2-144
0.007(6.46)	0.014(13.14)	6.489	8.309	2-145
0.007(6.46)	0.014(13.14)	6.509	8.289	2-146
0.007(6.46)	0.014(13.14)	6.464	8.333	2-147
0.007(6.46)	0.014(13.14)	6.484	8.314	2-148
0.007(6.46)	0.014(13.14)	6.503	8.295	2-149
0.007(6.46)	0.014(13.14)	6.522	8.277	2-150
0.007(6.46)	0.014(13.14)	6.498	8.300	2-151
0.007(6.46)	0.014(13.14)	6.564	8.236	2-152
0.007(6.46)	0.014(13.14)	6.622	8.180	2-153
0.007(6.46)	0.014(13.14)	6.576	8.224	2-154

Table 23. (Continued)

W (cross-sectional dia.) = .103[.003]

SQUEEZE, (%) MIN	SQUEEZE, (%) MAX	PERIPHERAL COMPRESSION, % MIN	PERIPHERAL COMPRESSION, % MAX	O-RING
0.007(6.46)	0.014(13.14)	6.626	8.176	2-155
0.007(6.46)	0.014(13.14)	6.628	8.174	2-156
0.007(6.46)	0.014(13.14)	6.670	8.133	2-157
0.007(6.46)	0.014(13.14)	6.708	8.097	2-158
0.007(6.46)	0.014(13.14)	6.651	8.152	2-159
0.007(6.46)	0.014(13.14)	6.686	8.118	2-160
0.007(6.46)	0.014(13.14)	6.718	8.087	2-161
0.007(6.46)	0.014(13.14)	6.747	8.059	2-162
0.007(6.46)	0.014(13.14)	6.774	8.032	2-163
0.007(6.46)	0.014(13.14)	6.726	8.079	2-164
0.007(6.46)	0.014(13.14)	6.752	8.054	2-165
0.007(6.46)	0.014(13.14)	6.775	8.031	2-166
0.007(6.46)	0.014(13.14)	6.797	8.010	2-167
0.007(6.46)	0.014(13.14)	6.755	8.051	2-168
0.007(6.46)	0.014(13.14)	6.776	8.030	2-169
0.007(6.46)	0.014(13.14)	6.796	8.011	2-170
0.007(6.46)	0.014(13.14)	6.815	7.992	2-171
0.007(6.46)	0.014(13.14)	6.777	8.029	2-172
0.007(6.46)	0.014(13.14)	6.795	8.012	2-173
0.007(6.46)	0.014(13.14)	6.813	7.995	2-174
0.007(6.46)	0.014(13.14)	6.829	7.979	2-175
0.007(6.46)	0.014(13.14)	6.795	8.012	2-176
0.007(6.46)	0.014(13.14)	6.811	7.997	2-177
0.007(6.46)	0.014(13.14)	6.826	7.982	2-178

W (cross-sectional dia.) = .139[.004]

0.008(5.92)	0.018(12.21)	5.630	9.123	2-210
0.008(5.92)	0.018(12.21)	5.734	9.025	2-211
0.008(5.92)	0.018(12.21)	5.829	8.937	2-212
0.008(5.92)	0.018(12.21)	5.911	8.859	2-213
0.008(5.92)	0.018(12.21)	5.987	8.788	2-214
0.008(5.92)	0.018(12.21)	6.055	8.724	2-215
0.008(5.92)	0.018(12.21)	5.980	8.795	2-216
0.008(5.92)	0.018(12.21)	6.042	8.736	2-217
0.008(5.92)	0.018(12.21)	6.099	8.681	2-218
0.008(5.92)	0.018(12.21)	6.151	8.632	2-219
0.008(5.92)	0.018(12.21)	6.200	8.585	2-220
0.008(5.92)	0.018(12.21)	6.245	8.543	2-221
0.008(5.92)	0.018(12.21)	6.125	8.657	2-222
0.008(5.92)	0.018(12.21)	6.211	8.575	2-223

Table 23.   (Continued)

| SQUEEZE,(%) | | PERIPHERAL COMPRESSION,% | | O-RING |
MIN	MAX	MIN	MAX	
0.008(5.92)	0.018(12.21)	6.286	8.503	2-224
0.008(5.92)	0.018(12.21)	6.220	8.567	2-225
0.008(5.92)	0.018(12.21)	6.286	8.503	2-226
0.008(5.92)	0.018(12.21)	6.345	8.447	2-227
0.008(5.92)	0.018(12.21)	6.323	8.467	2-228
0.008(5.92)	0.018(12.21)	6.375	8.418	2-229
0.008(5.92)	0.018(12.21)	6.423	8.373	2-230
0.008(5.92)	0.018(12.21)	6.466	8.331	2-231
0.008(5.92)	0.018(12.21)	6.380	8.414	2-232
0.008(5.92)	0.018(12.21)	6.421	8.374	2-233
0.008(5.92)	0.018(12.21)	6.459	8.337	2-234
0.008(5.92)	0.018(12.21)	6.494	8.303	2-235
0.008(5.92)	0.018(12.21)	6.527	8.272	2-236
0.008(5.92)	0.018(12.21)	6.558	8.242	2-237
0.008(5.92)	0.018(12.21)	6.586	8.215	2-238
0.008(5.92)	0.018(12.21)	6.516	8.283	2-239
0.008(5.92)	0.018(12.21)	6.544	8.256	2-240
0.008(5.92)	0.018(12.21)	6.570	8.230	2-241
0.008(5.92)	0.018(12.21)	6.595	8.206	2-242
0.008(5.92)	0.018(12.21)	6.618	8.184	2-243
0.008(5.92)	0.018(12.21)	6.599	8.203	2-244
0.008(5.92)	0.018(12.21)	6.621	8.181	2-245
0.008(5.92)	0.018(12.21)	6.641	8.161	2-246
0.008(5.92)	0.018(12.21)	6.661	8.142	2-247
0.008(5.92)	0.018(12.21)	6.680	8.124	2-248
0.008(5.92)	0.018(12.21)	6.606	8.195	2-249
0.008(5.92)	0.018(12.21)	6.612	8.190	2-250
0.008(5.92)	0.018(12.21)	6.644	8.159	2-251
0.008(5.92)	0.018(12.21)	6.661	8.142	2-252
0.008(5.92)	0.018(12.21)	6.678	8.126	2-253
0.008(5.92)	0.018(12.21)	6.694	8.111	2-254
0.008(5.92)	0.018(12.21)	6.709	8.096	2-255
0.008(5.92)	0.018(12.21)	6.724	8.081	2-256
0.008(5.92)	0.018(12.21)	6.738	8.068	2-257
0.008(5.92)	0.018(12.21)	6.751	8.055	2-258
0.008(5.92)	0.018(12.21)	6.704	8.100	2-259
0.008(5.92)	0.018(12.21)	6.731	8.075	2-260
0.008(5.92)	0.018(12.21)	6.755	8.051	2-261
0.008(5.92)	0.018(12.21)	6.777	8.029	2-262
0.008(5.92)	0.018(12.21)	6.736	8.069	2-263
0.008(5.92)	0.018(12.21)	6.758	8.048	2-264
0.008(5.92)	0.018(12.21)	6.778	8.028	2-265
0.008(5.92)	0.018(12.21)	6.797	8.010	2-266

Table 23. (Continued)

W (cross-sectional dia.) = .139[.004]

SQUEEZE,(%)		PERIPHERAL COMPRESSION,%		O-RING
MIN	MAX	MIN	MAX	
0.008(5.92)	0.018(12.21)	6.760	8.046	2-267
0.008(5.92)	0.018(12.21)	6.779	8.028	2-268
0.008(5.92)	0.018(12.21)	6.796	8.011	2-269
0.008(5.92)	0.018(12.21)	6.813	7.994	2-270
0.008(5.92)	0.018(12.21)	6.779	8.027	2-271
0.008(5.92)	0.018(12.21)	6.796	8.011	2-272
0.008(5.92)	0.018(12.21)	6.811	7.996	2-273
0.008(5.92)	0.018(12.21)	6.826	7.982	2-274

W (cross-sectional dia.) = .210[.005]

0.012(5.85)	0.024(10.91)	5.763	8.998	2-315
0.012(5.85)	0.024(10.91)	5.846	8.920	2-316
0.012(5.85)	0.024(10.91)	5.920	8.851	2-317
0.012(5.85)	0.024(10.91)	5.988	8.787	2-318
0.012(5.85)	0.024(10.91)	6.049	8.729	2-319
0.012(5.85)	0.024(10.91)	5.981	8.793	2-320
0.012(5.85)	0.024(10.91)	6.038	8.740	2-321
0.012(5.85)	0.024(10.91)	6.091	8.689	2-322
0.012(5.85)	0.024(10.91)	6.139	8.643	2-323
0.012(5.85)	0.024(10.91)	6.185	8.600	2-324
0.012(5.85)	0.024(10.91)	6.116	8.665	2-325
0.012(5.85)	0.024(10.91)	6.197	8.588	2-326
0.012(5.85)	0.024(10.91)	6.268	8.520	2-327
0.012(5.85)	0.024(10.91)	6.332	8.460	2-328
0.012(5.85)	0.024(10.91)	6.270	8.519	2-329
0.012(5.85)	0.024(10.91)	6.327	8.464	2-330
0.012(5.85)	0.024(10.91)	6.379	8.415	2-331
0.012(5.85)	0.024(10.91)	6.425	8.370	2-332
0.012(5.85)	0.024(10.91)	6.403	8.391	2-333
0.012(5.85)	0.024(10.91)	6.445	8.351	2-334
0.012(5.85)	0.024(10.91)	6.484	8.314	2-335
0.012(5.85)	0.024(10.91)	6.519	8.279	2-336
0.012(5.85)	0.024(10.91)	6.441	8.355	2-337
0.012(5.85)	0.024(10.91)	6.476	8.321	2-338
0.012(5.85)	0.024(10.91)	6.508	8.290	2-339
0.012(5.85)	0.024(10.91)	6.538	8.261	2-340
0.012(5.85)	0.024(10.91)	6.566	8.234	2-341
0.012(5.85)	0.024(10.91)	6.499	8.299	2-342
0.012(5.85)	0.024(10.91)	6.526	8.272	2-343

Table 23. (Continued)

W (cross-sectional dia.) = .210[.005]

SQUEEZE,(%)		PERIPHERAL COMPRESSION,%		O-RING
MIN	MAX	MIN	MAX	
0.012(5.85)	0.024(10.91)	6.552	8.247	2-344
0.012(5.85)	0.024(10.91)	6.577	8.224	2-345
0.012(5.85)	0.024(10.91)	6.600	8.201	2-346
0.012(5.85)	0.024(10.91)	6.581	8.219	2-347
0.012(5.85)	0.024(10.91)	6.603	8.198	2-348
0.012(5.85)	0.024(10.91)	6.624	8.178	2-349
0.012(5.85)	0.024(10.91)	6.644	8.159	2-350
0.012(5.85)	0.024(10.91)	6.662	8.141	2-351
0.012(5.85)	0.024(10.91)	6.680	8.124	2-352
0.012(5.85)	0.024(10.91)	6.575	8.225	2-353
0.012(5.85)	0.024(10.91)	6.594	8.207	2-354
0.012(5.85)	0.024(10.91)	6.612	8.190	2-355
0.012(5.85)	0.024(10.91)	6.630	8.173	2-356
0.012(5.85)	0.024(10.91)	6.646	8.157	2-357
0.012(5.85)	0.024(10.91)	6.662	8.141	2-358
0.012(5.85)	0.024(10.91)	6.677	8.126	2-359
0.012(5.85)	0.024(10.91)	6.692	8.112	2-360
0.012(5.85)	0.024(10.91)	6.706	8.098	2-361
0.012(5.85)	0.024(10.91)	6.690	8.114	2-362
0.012(5.85)	0.024(10.91)	6.716	8.088	2-363
0.012(5.85)	0.024(10.91)	6.741	8.065	2-364
0.012(5.85)	0.024(10.91)	6.763	8.043	2-365
0.012(5.85)	0.024(10.91)	6.723	8.082	2-366
0.012(5.85)	0.024(10.91)	6.745	8.061	2-367
0.012(5.85)	0.024(10.91)	6.766	8.041	2-368
0.012(5.85)	0.024(10.91)	6.785	8.022	2-369
0.012(5.85)	0.024(10.91)	6.745	8.052	2-370
0.012(5.85)	0.024(10.91)	6.767	8.039	2-371
0.012(5.85)	0.024(10.91)	6.785	8.022	2-372
0.012(5.85)	0.024(10.91)	6.802	8.006	2-373
0.012(5.85)	0.024(10.91)	6.769	8.038	2-374
0.012(5.85)	0.024(10.91)	6.785	8.022	2-375
0.012(5.85)	0.024(10.91)	6.800	8.007	2-376
0.012(5.85)	0.024(10.91)	6.815	7.992	2-377

Table 23. (Continued)
W (cross-sectional dia.) = .275[.006]

SQUEEZE,(%)		PERIPHERAL COMPRESSION,%		O-RING
MIN	MAX	MIN	MAX	
0.016(5.79)	0.030(10.34)	6.551	8.249	2-425
0.016(5.79)	0.030(10.34)	6.572	8.229	2-426
0.016(5.79)	0.030(10.34)	6.592	8.209	2-427
0.016(5.79)	0.030(10.34)	6.611	8.191	2-428
0.016(5.79)	0.030(10.34)	6.561	8.239	2-429
0.016(5.79)	0.030(10.34)	6.580	8.221	2-430
0.016(5.79)	0.030(10.34)	6.598	8.204	2-431
0.016(5.79)	0.030(10.34)	6.615	8.187	2-432
0.016(5.79)	0.030(10.34)	6.631	8.171	2-433
0.016(5.79)	0.030(10.34)	6.647	8.155	2-434
0.016(5.79)	0.030(10.34)	6.663	8.141	2-435
0.016(5.79)	0.030(10.34)	6.677	8.126	2-436
0.016(5.79)	0.030(10.34)	6.691	8.113	2-437
0.016(5.79)	0.030(10.34)	6.676	8.127	2-438
0.016(5.79)	0.030(10.34)	6.703	8.102	2-439
0.016(5.79)	0.030(10.34)	6.727	8.078	2-440
0.016(5.79)	0.030(10.34)	6.750	8.056	2-441
0.016(5.79)	0.030(10.34)	6.711	8.094	2-442
0.016(5.79)	0.030(10.34)	6.732	8.073	2-443
0.016(5.79)	0.030(10.34)	6.753	8.053	2-444
0.016(5.79)	0.030(10.34)	6.772	8.034	2-445
0.016(5.79)	0.030(10.34)	6.704	8.101	2-446
0.016(5.79)	0.030(10.34)	6.741	8.065	2-447
0.016(5.79)	0.030(10.34)	6.774	8.032	2-448
0.016(5.79)	0.030(10.34)	6.805	8.003	2-449

Design Example 13 Design of Rotary O-Ring Seal Using Design Tables (Refer to Fig. 35)

Design a floating shaft seal for a shaft approximately 2.5 in. in diameter, running at 700 rpm and sealing a pressure of 900 psi. *Design procedure*:

1. Shaft speed:

$$(700 \text{ rpm})(2.5 \text{ in.}) \frac{\pi}{12 \text{ in./ft}} = 485 \text{ fpm}$$

2. O-ring selection: According to Table 21, either a 0.070-in.- or 0.103-in.-cross-sectional-diameter O-ring can be used since the shaft speed is 485 fpm. As a general rule, the smallest O-ring cross section should be chosen to minimize the dynamic friction at the O-ring/shaft interface. Therefore, select the 0.070-in.-diameter-cross-section O-ring of size 2-039 for a shaft 2.539 in. in diameter.

3. Dimensions:

   From Table 21:

   Width of O-ring groove = 0.079 ± 0.001 in.

   From Table 22:

   Shaft diameter = 2.5390 ± 0.0005 in. diam.

   Groove diameter = 2.666 ± 0.001 in. diam.

   Housing bore = 2.5450 ± 0.0005 in. diam.

4. Check on the peripheral compression:

   $$\text{Peripheral compression} = \frac{(2.879 \pm 0.026) - (2.666 \pm 0.001)}{2.879 \pm 0.026}$$

   $$\text{Minimum peripheral compression} = \frac{2.853 - 2.667}{2.853} \times 100\% = 6.5\%$$

   $$\text{Maximum peripheral compression} = \frac{2.905 - 2.665}{2.905} \times 100\% = 8.3\%$$

These values are given in Table 23.

5. Check on the O-ring cross-sectional squeeze: For a nominal 7% peripheral compression we can use a 2.2% increase in the O-ring cross section according to Fig. 23. Then the installed O-ring will have a cross section of

   Minimum = (0.070 − 0.003)(1.022) = 0.0685 in.

   Maximum = (0.070 + 0.003)(1.022) = 0.0746 in.

The O-ring cross-sectional squeeze when installed within the gland is

$$\text{Minimum} = 0.0685 - \frac{2.667 - 2.5385}{2} = 0.0042$$

$$= \frac{0.0042}{0.0685} \times 100\% = 6.17\%$$

$$\text{Minimum} = 0.0746 - \frac{2.665 - 2.5395}{2} = 0.0119$$

$$= \frac{0.0119}{0.0746} \times 100\% = 15.89\%$$

These values are given in Table 23.

APPENDIX 6A. DETERMINATION OF RELATIONSHIP BETWEEN PERCENT OF DIAMETRAL REDUCTION OF O-RING AND PERCENT INCREASE IN CROSS-SECTIONAL WIDTH (Fig. 23)

Consider an O-ring being installed into an O-ring groove of smaller diameter

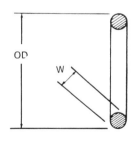
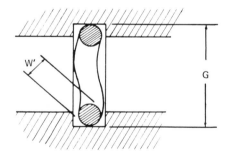

$$\% \text{ of diametral reduction} = \left(\frac{OD - G}{OD}\right) \times 100\% = \frac{d(OD)}{OD} \times 100\%$$

$$\% \text{ increase in cross-sectional width} = \left(\frac{W' - W}{W}\right) \times 100\%$$

$$= \frac{dW}{W} \times 100\%$$

Relationships:

1. Conservation of mass (if cross section is considered to be always circular):

$$\frac{\pi W^2}{4}(OD - W)\pi = \frac{\pi (W')^2}{4}(G - W')\pi$$

therefore

$$W^2(OD) - W^3 = (W')^2 G - (W')^3$$

$$W^2(OD) = (W')^2 G + W^3 - (W')^3$$

2. Since OD and G differ much more than W and W' and since W^3 is very close to $(W')^3$, we can ignore the last two cubed terms. They are of the order 0.0003, compared to OD and G, which are of an order greater than 0.210. For example, if we inserted typical values:

$$W^2(OD) = (W')^2 G + W^3 - (W')^3$$

$$(0.070)^2(2.372) \approx (0.075)^2(2.196) + (0.070)^3 - (0.075)^3$$

$$0.0049(2.372) \approx 0.0056(2.196) + 0.00034 - 0.00042$$

$$0.0116 \approx 0.0123 - 0.00008$$

The last term of the equation is small compared to the other terms and therefore can be ignored, giving

$$W^2(OD) = (W')^2 G$$

This states the square of O-ring width times the diameter is a constant

$$W^2(D) = K$$

Take the derivative

$$2W\, dW(D) + W^2\, d(D) = 0$$

therefore

$$2W\, dW = -W^2\, d(D)/D$$

$$2\frac{dW}{W} = \frac{-d(D)}{D}$$

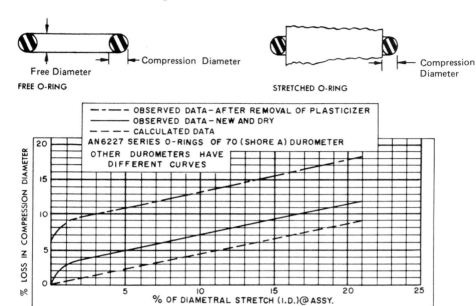

Figure 36. Loss in compression diameter due to stretch. In the stretched condition, an O-ring cross section is no longer circular. It is often necessary to compensate for the loss in squeeze resulting from the reduced compression diameter. Dimensional changes in the free diameter do not affect the seal. (From *O-Ring Handbook*, Parker Seal Co., Lexington, Ky., October 1967)

3. Therefore (with cross sections always circular)

$$\frac{\% \text{ increase in W}}{\% \text{ decrease in D}} = \frac{dW/W}{-d(D)/D} = \frac{1}{2}$$

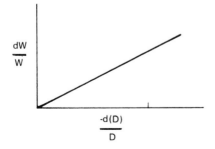

4. Because the O-ring cross section does not remain circular when
the O-ring is pressed into the smaller groove, the minimum width
W' will be less than that calculated through Step 3. Thus, the
actual W' can be estimated to be about 2.5 percent less using a
logical analogy from Fig. 36. Therefore, the actual data would
probably fall between the calculated and estimated curves shown
below.

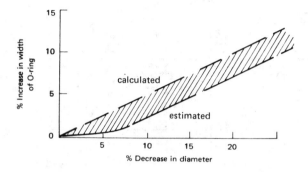

References

1. *O-Ring Design and Selection Handbook*, No. 110-A, Stillman Seal Division, Sargent Industries, Carlsbad, Calif., 1976.
2. *O-Ring Handbook OR5700*, Parker Seal Company, Lexington, Ky., January 1977.
3. *Parker O-Ring Handbook*, Parker Seal Company, Culver City, Calif., 1968.
4. *Seal Compound Manual*, Parker Seal Company, Culver City, Calif., 1964, pp. 1-2.
5. Richard S. Fein, Boundary Lubrication, *Lubrication* (Texaco, Inc.) 57(1):3-12 (1971).
6. O-ring insertion tool developed by L. J. Martini, Naval Ocean Systems Center, Patent Application.

Index

DATE DUE